SpringerBriefs in Applied Sciences and Technology

SpringerBriefs present concise summaries of cutting-edge research and practical applications across a wide spectrum of fields. Featuring compact volumes of 50 to 125 pages, the series covers a range of content from professional to academic.

Typical publications can be:

- A timely report of state-of-the art methods
- An introduction to or a manual for the application of mathematical or computer techniques
- A bridge between new research results, as published in journal articles
- A snapshot of a hot or emerging topic
- An in-depth case study
- A presentation of core concepts that students must understand in order to make independent contributions.

SpringerBriefs are characterized by fast, global electronic dissemination, standard publishing contracts, standardized manuscript preparation and formatting guidelines, and expedited production schedules.

On the one hand, **SpringerBriefs in Applied Sciences and Technology** are devoted to the publication of fundamentals and applications within the different classical engineering disciplines as well as in interdisciplinary fields that recently emerged between these areas. On the other hand, as the boundary separating fundamental research and applied technology is more and more dissolving, this series is particularly open to trans-disciplinary topics between fundamental science and engineering.

Indexed by EI-Compendex, SCOPUS and Springerlink.

Katarzyna Chojnacka · Filip Gil ·
Dawid Skrzypczak · Grzegorz Izydorczyk

Sustainable Fertilizers: Utilizing Biomass Ashes and Biochars

Transforming Waste into Agricultural Products

Springer

Katarzyna Chojnacka
Department of Advanced Material Technologies
Wrocław University of Science and Technology
Wrocław, Poland

Filip Gil
Department of Advanced Material Technologies
Wrocław University of Science and Technology
Wrocław, Poland

Dawid Skrzypczak
Department of Advanced Material Technologies
Wrocław University of Science and Technology
Wrocław, Poland

Grzegorz Izydorczyk
Department of Advanced Material Technologies
Wrocław University of Science and Technology
Wrocław, Poland

ISSN 2191-530X ISSN 2191-5318 (electronic)
SpringerBriefs in Applied Sciences and Technology
ISBN 978-3-032-03764-0 ISBN 978-3-032-03765-7 (eBook)
https://doi.org/10.1007/978-3-032-03765-7

This Springer imprint is published by the registered company Springer Nature Switzerland AG
The registered company address is: Gewerbestrasse 11, 6330 Cham, Switzerland

Preface

This book provides an in-depth examination of the potential use of biomass ash and biochar as sustainable fertilizers and soil conditioners within the context of circular and climate-resilient agriculture. In light of increasing concerns over soil degradation, environmental contamination, and the depletion of finite mineral fertilizer resources, it evaluates biomass-based materials as viable alternatives that support nutrient cycling and reduce dependency on synthetic agrochemicals. Biomass ash, generated from the combustion of plant residues, is notable for its high content of essential macronutrients (K, Ca, P, and Mg) and strong alkalinity, making it effective in soil pH correction and nutrient replenishment. Biochar, obtained through pyrolysis under limited oxygen conditions, functions primarily as a stable, porous carbon amendment that improves soil structure, enhances microbial activity, and contributes to long-term carbon sequestration. The book systematically explores the origins, characteristics, production techniques, and agronomic effects of these materials, drawing upon current scientific research and field data. Special emphasis is placed on compositional variability, environmental safety (heavy metals and organic pollutants), and the legal and regulatory frameworks that govern their agricultural application, including recent European Union directives. Through a multidisciplinary perspective, the volume highlights the role of biomass ash and biochar in enhancing soil fertility, supporting plant productivity, and promoting ecological sustainability in agricultural systems. This publication offers valuable insights and practical guidance for researchers, policymakers, and agricultural practitioners working toward sustainable land management and the development of bio-based fertilization strategies.

Wrocław, Poland

Katarzyna Chojnacka
Filip Gil
Dawid Skrzypczak
Grzegorz Izydorczyk

Competing Interests The authors have no competing interests to declare that are relevant to the content of this manuscript.

Contents

Abbreviations

ABP	Animal by-products
CDR	Carbon Dioxide Removal
CEC	Cation Exchange Capacity
CMC	Component Material Category
CRCF	Carbon Removal Certification Framework
DM	Dry matter
EBC	European Biochar Certificate
EPA	Environmental Protection Agency
EU	European Union
FPR	Fertilizing Products Regulation
GWP	Global Warming Potential
IBI	International Biochar Initiative
IPCC	Intergovernmental Panel on Climate Change
ISO	International Organization for Standardization
LCA	Life Cycle Assessment
NPK	Nitrogen, Phosphorus, Potassium
PAHs	Polycyclic Aromatic Hydrocarbons
PCDD/F	Polychlorinated Dibenzo-p-Dioxins and Furans
PFC	Product Functional Category
SOC	Soil Organic Carbon
TRL	Technology Readiness Level
WHO-TEQ	World Health Organization Toxic Equivalents

Chapter 1
Introduction to Precision Agriculture

1.1 Purpose and Scope of the Book

Agriculture today faces the dual challenge of ensuring sufficient food production for a growing population while preserving the environment. The intensification of agricultural production in the twentieth century, based on artificial fertilizers and agrochemicals, has significantly increased yields, but has also caused a number of environmental problems: soil degradation, water pollution, greenhouse gas emissions, and loss of biodiversity [1]. In the face of these threats, the concept of sustainable agriculture has gained crucial importance—it is considered the foundation for ensuring food security and improving the quality of life of societies [2]. Sustainable agriculture seeks to manage production in a way that increases yields in an environmentally friendly manner, reduces ecosystem degradation, and conserves resources for future generations [3]. Achieving these goals, however, requires innovative solutions, including alternative fertilizers and soil improvers that maintain or increase soil fertility and yields with less chemical pressure on the environment [4].

This book is devoted to just such solutions—the use of biomass ash and biochar as fertilizers and soil improvers in agriculture. Biomass ash (e.g., ash generated from burning wood or straw) is rich in macronutrients (calcium, potassium, phosphorus, and magnesium) and exhibits deacidifying effects, making it a potential replacement for agricultural lime and conventional mineral fertilizers [5]. On the other hand, biochar (biochar)—a highly carbonized material obtained by pyrolysis of biomass—is characterized by an exceptionally high content of stable organic carbon. As a result, it can act as a long-term carbon store in the soil, contributing to CO_2 sequestration, while improving the physicochemical properties of soils (including increasing water and sorption capacity) [6]. In recent years, biochar has been seen as a promising soil additive with the potential to raise soil fertility and yields while reducing greenhouse gas emissions. Biomass ash, on the other hand, fits the model of a closed-loop economy—instead of being treated as waste, it can be returned to the matter cycle as a fertilizer or soil improver [7].

K. Chojnacka et al., *Sustainable Fertilizers: Utilizing Biomass Ashes and Biochars*, SpringerBriefs in Applied Sciences and Technology,
https://doi.org/10.1007/978-3-032-03765-7_1

The aim of the book is to present the current scientific knowledge of both these materials—ash and biochar—and to assess their usefulness in sustainable agriculture. The chapters discuss the sources and properties of biomass ash and biochar, their effects on soil properties (physical, chemical and biological), on yield and quality of agricultural crops, as well as on the environment (gaseous emissions, carbon sequestration, and pollution risks). Special emphasis is placed on the latest research results (both field experiments and laboratory analysis and modeling) and the place of these solutions in the concepts of circular agriculture and climate change mitigation. The book is addressed to a wide audience—scientists, agricultural advisors, policy makers, and practitioners interested in implementing sustainable agriculture principles. It is hoped that this publication will facilitate understanding of the potential of biomass ash and biochar management, but also the limitations associated with it. Indeed, it should be emphasized that the quality of these materials and appropriate methods of application are key. Ashes may contain undesirable admixtures (e.g., heavy metals) that preclude their direct agricultural use without prior preparation or quality control [5]. Biochar, on the other hand, exhibits high variability in properties depending on the raw material and production parameters, which affects its performance in the soil. The following chapters examine these issues in detail, providing both research results and practical guidance for the safe and effective use of ash and biochar in agriculture.

1.2 The Importance of Sustainable Agriculture

Sustainable agriculture now plays a key role in global development strategies. This is due to the fact that agriculture is the basis of food security, while at the same time it has a strong impact on the environment. The traditional model of intensive agriculture of the "green revolution" allowed to increase food production, but this was often at the expense of degradation of natural resources. Today, agriculture is responsible for a significant portion of greenhouse gas emissions, contributes to deforestation, loss of biodiversity and pollution of soils and waters through excessive use of nitrogen fertilizers and pesticides [8]. Such a model is unsustainable in the long run, especially with population projections rising to about 10 billion by the middle of the twenty-first century. Sustainable agriculture appears as a necessary response to these challenges—an approach that combines production goals with the requirements of protecting the environment and natural resources for future generations [9, 10].

The importance of sustainable agriculture is underscored by global initiatives and agreements. As early as the 1987 Brundtland Report, it was noted that development (including agricultural development) should meet the needs of the current generation without detracting from the opportunities of future generations. The UN's 2030 Agenda has made sustainable agriculture an essential component of several Sustainable Development Goals, including eradicating hunger (Goal 2), protecting water and soils (Goals 14 and 15) and climate action (Goal 13). Indeed, sustainable agriculture has the unique potential to contribute to solving hunger while mitigating climate change and protecting ecosystems [11]. For example, practices such as regenerative agriculture, agroforestry, organic farming, and integrated pest management help increase soil fertility and crop productivity in a nature-friendly way. They improve water and organic matter retention in soils, reduce erosion and reduce the need for chemical agroinputs.

Moreover, sustainable agriculture is closely linked to social and economic development in rural areas. Research indicates that implementing sustainable practices on the land promotes the long-term economic viability of farms, reducing costs (e.g., by recycling nutrients within the farm) and increasing resilience to change (e.g., by improving soil health) [2]. Thus, sustainable agriculture contributes to the quality of life of rural residents, the stability of local communities and the food security of entire countries [2].

In the context of sustainable development, there is growing interest in alternative fertilizers and bioproducts, which make it possible to close the cycle of matter in agriculture and reduce dependence on non-renewable raw materials. An excellent example of this is precisely the fertilizers made from biomass waste, such as ash or biochar, discussed in this book. They fit in with the idea of circular agriculture, where residues and by-products (e.g., ash from biomass combustion) are returned to the soil as a source of nutrients, instead of becoming a nuisance waste [7]. This approach has multifaceted benefits: on the one hand, it increases the self-sufficiency and resilience (resilience) of agricultural systems, and on the other hand, it reduces the pressure of agriculture on the environment by reducing waste and emissions. The following introductory sections take a closer look at why biomass ash and biochar are seen as vital pieces of this agricultural sustainability puzzle [12–14].

1.3 History and Development of Biomass Ash Fertilizers and Biochar

Wood ash is one of the oldest fertilizers used by humans [15]. Already in traditional agriculture, ashes from campfires and hearths have been used for thousands of years to fertilize fields—they were added to soils to increase fertility in mineral-poor soils. Historical records indicate that wood ashes were valued as a source of potash (potassium carbonate) and calcium—the name potash in English comes precisely from the ash held in a pot (pot ash) to extract potash [16]. In pre-industrial Europe and colonial America, wood ash was an important commodity, used not only in agriculture, but also in the production of soap, glass and gunpowder [17]. It was not until the nineteenth century that the discovery of rich deposits of potash salts and the development of the chemical industry meant that the traditional extraction of potash from ashes lost its importance on an industrial scale—potash fertilizers began to be made from fossil raw materials on a mass scale. Nevertheless, at the household and local level, wood ash continued to be used for decades as a fertilizer in fields and gardens, especially where it was readily available [16].

In the second half of the twentieth century, interest in biomass ash as a fertilizer was revived in a somewhat different context. The development of biomass-based power generation (burning waste wood, straw, etc. for heating and energy purposes) resulted in the generation of large quantities of ash as a by-product. Initially, they were treated as waste requiring landfilling. However, as early as the 1970s, experiments began in Scandinavia with the reintroduction of ash into forest ecosystems—for example, in Finland and Sweden, the liming of acid forest soils with wood ash was carried out on a massive scale to compensate for the loss of components carried away from biomass [18]. Over time, these experiments were also transferred to agriculture. As part of the idea of a closed-loop economy, an approach emerged to treat ash from biomass CHP plants as fertilizer returning to agricultural fields [19]. In many countries (including Germany, Austria, and the Nordic countries), quality standards and guidelines have been developed to regulate the agricultural use of ash—so as to benefit from its fertilizer value while minimizing environmental risks. This is because wood ash contains significant amounts of nutrients (in some ashes even approx. 20% of P_2O_5, 50% of K_2O and more than 90% of CaO in terms of pure fertilizer constituents), making it an attractive substitute for phosphate-potassium fertilizers and lime [7, 18]. On the other hand, it may contain contaminants—especially heavy metals (Cd, Pb, Ni, Hg) or dioxins—from contaminated raw materials or suboptimal combustion conditions [20, 21]. Therefore, when implementing ash for fertilization, it is necessary to comply with purity standards. Modern research indicates that many biomass ashes meet the criteria for approval for fertilizer use, especially ashes from the combustion of clean wood or straw. In cases of more contaminated ashes, methods of improving them (e.g., granulation, leaching of undesirable components) are being developed for safe management. Overall, the development of the use of biomass ash as a fertilizer has gained momentum in the twenty-first century, part of a broader trend of finding more sustainable sources of fertilizer and recycling nutrients [22, 23].

Biochar has an equally long, if not longer, history of use in agriculture, although its name and scientific interest in it have only recently emerged. Historically, mankind has long known the beneficial effects of charred organic residues on soil fertility. Archaeological evidence from the Amazon region indicates that as early as thousands of years ago, pre-Columbian communities there deliberately added charcoal and organic matter to soils, creating so-called terra preta—"black earths" characterized by extremely high fertility [24]. These anthropogenic black soils have survived centuries of intensive use in humid tropical climates, confirming the long-term positive effect of charcoal in soils [25]. In other parts of the world, too, we find examples of the traditional practice of adding charred biomass to the soil—whether in the form of residues from burning forests for cultivation (shifting cultivation) or carbon-rich wood ash. Ancient farmers from various civilizations (e.g., in Asia or the Middle East) observed that soils enriched in carbonaceous matter retained moisture better and produced higher yields [25].

Despite this past experience, modern science did not rediscover biochar until the twenty-first century. Earlier, in the nineteenth and twentieth centuries, soil scientists (like Justus von Liebig) focused on the importance of humus and organic matter for fertility, but the role of the stable carbon fraction was not separately analyzed

[26]. The breakthrough came in the 1990s, when research on terra preta brought the extraordinary properties of these soils to the attention of the scientific world. This was followed by experiments with the production of biochar—intentionally produced charcoal for soil use—and studies of its effects on soil properties and plants. The oil crisis of the 1970s accelerated the development of biomass pyrolysis technology, which indirectly laid the foundation for modern industrial-scale production of biochar [4]. As early as the 1980s and 1990s, the first commercial pyrolysis reactors were developed to obtain charcoal (and bio-oil and gas) from various organic wastes [4]. However, it was the beginning of the twenty-first century that saw a real blossoming of interest in biochar in the context of agriculture and climate protection. Several groundbreaking scientific papers published around 2005–2010 showed that the addition of biochar to soil could increase soil fertility, sequester carbon for hundreds of years and reduce nitrous oxide (a potent greenhouse gas) emissions from soil. These reports have generated tremendous interest both in the scientific community and among policymakers and practitioners' agriculture. Biochar began to be seen as a win–win technology, combining benefits for agricultural production with climate benefits [27]. The Intergovernmental Panel on Climate Change (IPCC) officially listed biochar as one of the negative emissions (CO_2 removal technologies) in climate stabilization scenarios [28]. Governments in some countries (including the UK and Japan) have included biochar in their low-carbon agricultural development strategies. The commercial sector is also growing dynamically—companies specializing in the production and application of biochar are being established, and in the voluntary carbon credit markets biochar already represents a significant position as a form of sustainable removal of carbon dioxide from the atmosphere.

In the scientific sphere, the development of biochar research has been exponential. At the beginning of the twenty-first century, only a single article per year was published on the subject, while today the number of publications reaches several thousand per year. It is estimated that over the past two decades, more than 12,000 scientific papers have been published on biochar, covering topics ranging from biochar production and modification methods, to its effects on crop soils, emissions and yields, to environmental applications (wastewater treatment, remediation of contaminated soils, etc.) [24]. Such tremendous interest translates into rapid advances in knowledge. We already know the basic mechanisms of action of biochar in soil—we know that it improves structure and water retention, increases the sorptive capacity (CEC) of soil, can stimulate the growth of soil microorganisms and bind heavy metals. At the same time, potential risks or limitations have been identified, such as the ability of biochar to adsorb nitrogen fertilizers (which sometimes results in a temporary decrease in nitrogen availability to plants) or the risk of introducing contaminants if the starting raw material was contaminated. Current research focuses, among other things, on optimizing the production of biochar (so that it has desirable properties, such as nutrient enrichment), the long-term effects of its use under different soil and climate conditions, and the full life cycle assessment (LCA) of these technologies [29, 30].

In summary, both biomass ash and biochar have come a long way from the traditional, intuitive soil improvers of old, through a period of oblivion in the fertilizer era, to their rediscovery in the twenty-first century as innovative tools for sustainable agriculture. Further chapters of the work detail the state of the art of these materials, showing their potential and limitations in the context of modern agricultural and environmental needs. In this way, the historical outline and current research developments form a coherent narrative: from the terra preta of the ancient Amazons, to advanced laboratory research and pilot implementations on twenty-first century farms—all confirming the value of drawing on the wisdom of nature and the cyclical cycle of matter to make agriculture more sustainable [24].

References

1. P. Sharma, P. Sharma, N. Thakur, Sustainable farming practices and soil health: a pathway to achieving SDGs and future prospects. Discov. Sustain. **5**(1), 1–14 (2024). https://doi.org/10.1007/S43621-024-00447-4/FIGURES/2
2. J. Polcyn, A. Stratan, V. Lopotenco, Sustainable agriculture's contribution to quality of life. Sustainability **15**(23), 16415 (2023). https://doi.org/10.3390/SU152316415
3. Y. Chen, Z. Sun, Y. Zhou, W. Yang, Y. Ma, The future of sustainable farming: an evolutionary game framework for the promotion of agricultural green production technologies. J. Clean. Prod. **17**, 142606 (2024). https://doi.org/10.1016/j.jclepro.2024.142606
4. N.L. Panwar, A. Pawar, B.L. Salvi, Comprehensive review on production and utilization of biochar. SN Appl. Sci. **1**(2), 1–19 (2019). https://doi.org/10.1007/S42452-019-0172-6/FIGURES/9
5. C. Jian et al., Effects of wood ash fertilizer on element dynamics in soil solution and crop uptake. Agronomy **15**(5), 1097 (2025). https://doi.org/10.3390/AGRONOMY15051097
6. C. Ji, K. Cheng, D. Nayak, G. Pan, Environmental and economic assessment of crop residue competitive utilization for biochar, briquette fuel and combined heat and power generation. J. Clean. Prod. **192**, 916–923 (2018)
7. E. Mayer et al., Utilization and recycling of wood ashes from industrial heat and power plants regarding fertilizer use. Waste Manage. **141**, 92–103 (2022). https://doi.org/10.1016/j.wasman.2022.01.027
8. A.O. Adewusi, O.F. Asuzu, T. Olorunsogo, C. Iwuanyanwu, E. Adaga, D.O. Daraojimba, AI in precision agriculture: a review of technologies for sustainable farming practices. World J. Adv. Res. Rev. **2024**(01), 2276–2285 (2024). https://doi.org/10.30574/wjarr.2024.21.1.0314
9. H. Knicker, R.E. White, The role of soil carbon sequestration as a climate change mitigation strategy: an Australian case study. Soil Syst. **6**(2), 46 (2022). https://doi.org/10.3390/SOILSYSTEMS6020046
10. E. Toensmeier, R. Ferguson, M. Mehra, Perennial vegetables: A neglected resource for biodiversity, carbon sequestration, and nutrition. PLoS ONE **15**(7) (2020). https://doi.org/10.1371/JOURNAL.PONE.0234611
11. R.N. Onwonga, *Carbon Sequestration: Pathway to Increased Agricultural Productivity and Zero Hunger for Developing Countries* (Springer, 2019), pp. 1–13. https://doi.org/10.1007/978-3-319-69626-3_60-1
12. D.H. Constance, Sustainable agriculture in the United States: a critical examination of a contested process. Sustainability **2**(1), 48–72 (2010). https://doi.org/10.3390/SU2010048
13. D. Paudel, R.C. Neupane, S. Sigdel, P. Poudel, A.R. Khanal, COVID-19 pandemic, climate change, and conflicts on agriculture: a trio of challenges to global food security. Sustainability (Switzerland) **15**(10) (2023). https://doi.org/10.3390/SU15108280
14. B.R.K. Runkle, Review: biological engineering for nature-based climate solutions. J. Biol. Eng. **16**(1) (2022). https://doi.org/10.1186/S13036-022-00287-8

15. S.C. Panday et al., Blending traditional knowledge of farmers in agriculture with modern scientific technologies in the hills of Uttarakhand state, in *Blending Indian Farmers' Traditional Knowledge in Agriculture with Modern Scientific Technologies* (2025), pp. 311–330. https://doi.org/10.1007/978-981-96-1020-4_14
16. U. Kleine-Kleffmann, The discovery of the first potash mine and the development of the potash industry since 1861. J. Plant Nutr. Soil Sci. **186**(6), 615–622 (2023). https://doi.org/10.1002/JPLN.202300382
17. O.F. Uzor, C.M. Ike, C.I. Nwankwo, Soap production using plantain peel ash as a source of alkali; a strategy for improving entrepreneurship education in tertiary institutions in Anambra State. ANACHEM J. **15**(1), 64–73 (2024). https://doi.org/10.5281/ZENODO.14567357
18. A. Demeyer, J.C. Voundi Nkana, M.G. Verloo, Characteristics of wood ash and influence on soil properties and nutrient uptake: an overview. Bioresour. Technol. **77**(3), 287–295 (2001). https://doi.org/10.1016/S0960-8524(00)00043-2
19. L. Prasittisopin, Power plant waste (fly ash, bottom ash, biomass ash) management for promoting circular economy in sustainable construction: emerging economy context. Smart Sustain. Built Environ. (2024). https://doi.org/10.1108/SASBE-09-2024-0395/FULL/PDF
20. X. Zhao, J. Yang, N. Ning, Z. Yang, Chemical stabilization of heavy metals in municipal solid waste incineration fly ash: a review. Environ. Sci. Pollut. Res. **29**(27), 40384–40402 (2022). https://doi.org/10.1007/S11356-022-19649-2
21. M.L. Ummik, O. Järvik, A. Konist, Dioxin concentrations and congener distribution in biomass ash from small to large scale biomass combustion plants. Environ. Sci. Pollut. Res. **31**(49), 58946–58956 (2024). https://doi.org/10.1007/S11356-024-35141-5/FIGURES/5
22. S.K. Das et al., A novel approach on leaching study for removal of toxic elements from thermal power plant-based fly ash using natural bio-surfactant. Case Stud. Chem. Environ. Eng. **4**, 100156 (2021). https://doi.org/10.1016/J.CSCEE.2021.100156
23. K. Buneviciene, D. Drapanauskaite, R. Mazeika, V. Tilvikiene, J. Baltrusaitis, Granulated biofuel ash as a sustainable source of plant nutrients. Waste Manage. Res. **39**(6), 806–817 (2021)
24. M. Varkolu et al., Recent advances in biochar production, characterization, and environmental applications. Catalysts **15**(3), 243 (2025). https://doi.org/10.3390/CATAL15030243
25. J. Lehmann, Terra preta Nova–where to from here, in *Amazonian Dark Earths: Wim Sombroek's Vision* (Springer, 2009). https://doi.org/10.1007/978-1-4020-9031-8.PDF
26. S.K. Gore, S.S. Rudrawar, Significance of rural development in national progress. https://doi.org/10.5281/ZENODO.10450592
27. S. Wang, L. Wang, S. Li, T. Zhang, K. Cai, The win–win effects of an invasive plant biochar on a soil–crop system: controlling a bacterial soilborne disease and stabilizing the soil microbial community network. Microorganisms **12**(3), 447 (2024). https://doi.org/10.3390/MICROORGANISMS12030447/S1
28. L. Chen et al., Biochar-augmented carbon-negative concrete. Chem. Eng. J. **431**, 133946 (2022). https://doi.org/10.1016/J.CEJ.2021.133946
29. J. Carvalho et al., Life cycle assessment (LCA) of biochar production from a circular economy perspective. Processes **10**(12), 2684 (2022). https://doi.org/10.3390/PR10122684
30. J. Matuštík, T. Hnátková, V. Kočí, Life cycle assessment of biochar-to-soil systems: a review. J. Clean. Prod. **259**, 120998 (2020). https://doi.org/10.1016/J.JCLEPRO.2020.120998

Chapter 2
Biomass Ash Definition and Types of Biomass Ash

2.1 Introduction

Biomass ash is the inorganic, non-combustible fraction that remains after the complete or partial oxidation of plant- or animal-derived biomass (e.g., wood, straw, poultry litter, sewage sludge) and is now viewed as a secondary nutrient source in circular agriculture. It represents the non-combustible mineral fraction and is increasingly viewed as a secondary nutrient source within circular agriculture frameworks [1]. The progressive substitution of fossil fuels by biomass has been accompanied by a steady rise in ash production. Recent energy-balance studies estimate global biomass ash output at approximately 40–60 Mt yr^{-1}, which is about one order of magnitude lower than earlier headline values [2]. Although management of this by-product poses logistical challenges, the ash represents a potential nutrient resource owing to its mineral content. Historically, ash derived from biomass combustion-wood ash in particular-served as one of the earliest mineral fertilizers. Agricultural application of ash enabled partial restitution of nutrients removed during crop harvest [3]. Contemporary circular-economy strategies advocate recycling of biomass ash to minimize disposal and close elemental loops [4].

Biomass ash comprises the non-combustible mineral fraction persisting after the organic matrix has been oxidized. In practice, this term covers ashes produced from the combustion of various types of biomass: from wood and forest waste, through straw and other agricultural residues, to biomass of animal origin (e.g., natural fertilizers, post-fermentation sludge) or sewage sludge [5, 6]. Biomass ashes are commonly categorized by origin: wood ash, agricultural ash (e.g., cereal straw, grain husks), animal biomass ash (e.g., poultry manure), and sewage sludge ash. Each type differs in ash yield, chemical composition, and agronomic potential [1]. Each of these groups is characterized by a different composition and properties.

As in the case of coal combustion, thermal processing of biomass usually produces two fractions of ash, depending on where they are formed in the combustion plant: bottom ash and fly ash. Bottom ash, also called furnace or grate ash, accumulates at the

K. Chojnacka et al., *Sustainable Fertilizers: Utilizing Biomass Ashes and Biochars*, SpringerBriefs in Applied Sciences and Technology,
https://doi.org/10.1007/978-3-032-03765-7_2

bottom of the combustion chamber and consists of heavier particles, slag, and sinter. Fly ash, consists of fine ash particles, carried along with the flue gas and then captured by dust removal devices (e.g., filters, electrostatic precipitators). Fly ash from biomass is usually characterized by smaller particle size and greater porosity, which makes it more chemically reactive. Studies also show some differences in the composition of both fractions. Volatile elements such as chlorine and heavy metals (e.g., Cd, Pb, and Zn) partially vaporize at high combustion temperatures and recondense on fine fly ash particles, resulting in higher concentrations in fly ash compared to bottom ash. Bottom ash generally contains significantly lower levels of these contaminants, a pattern reflected in regulatory restrictions adopted by several EU countries [7]. For this reason, some countries (e.g., Austria) have regulations prohibiting the use of biomass fly ash on soil to prevent excessive amounts of heavy metals from entering the environment. This type of ash is often sent to landfills as hazardous waste [7–9]. In this chapter the discussion of ash composition and suitability focuses on clean biomass sources, excluding cases such as coal or municipal-waste co-firing (wood, straw, etc.), without admixtures of fossil fuels, and omit specific cases such as co-firing with coal or municipal waste.

The ash content in biomass itself (i.e., the amount of ash produced after burning a given mass of fuel) varies greatly depending on the type of raw material. Wood biomass is a low-ash fuel. Clean wood leaves only about 0.5–3% ash in terms of dry mass. In comparison, biomass of agricultural origin (e.g., cereal straw, energy grasses) contains more minerals and produces more ash, in the range of 1–10% of dry mass [1]. Extreme values are observed in the case of animal or municipal waste biomass, e.g., dried animal manure or sewage sludge can leave up to 20–60% ash due to their high mineral content. Such high variability results from differences in plant anatomy and chemistry (wood consists mainly of carbon and oxygen, while straw contains more silica taken from the soil, etc.) and fuel impurities (e.g., sand, soil adhering to straw) [5, 6]. In practice, this means that biomass ash is not a homogeneous material; on the contrary, it exhibits significant heterogeneity compared to ash from coal combustion [1]. This feature complicates its standardization and utilization, but at the same time requires an individual approach to assessing the composition and properties of ash from different sources.

2.2 Production Process and Chemical Composition

The process of biomass ash formation is associated with the thermal conversion of fuel in power boilers, furnaces, or other combustion devices. During combustion, biomass undergoes complex processes of pyrolysis, oxidation, and melting of mineral residues. At temperatures of 800–1200 °C (typical for grate or fluidized bed boilers), most of the organic matter is burned into gaseous products (CO_2, H_2O, volatile compounds), while inorganic ash components (minerals and non-combustible elements) partially melt or remain in a solid phase, forming ash. The combustion method affects the properties of the ash, e.g., in boilers with lower temperatures, unburned carbon particles may remain (which increases the unburned content in the ash), and the cooling rate of the slag affects the grain structure (rapid quenching creates porous, glassy particles, while slow cooling promotes mineral crystallization). The combustion technology is also of great importance: in fluidized bed boilers, most of the ash remains in the furnace (bottom ash) due to the circulation of sand grains, while in pulverized coal boilers, a significant part forms fly ash carried away by the flue gas. Despite these variable factors, the primary determinant of ash composition is the original chemical composition of the biomass [10].

The chemical composition of biomass ash reflects the inorganic part of plant tissue or other combusted materials, concentrated after the removal of organic matter. The main macronutrients in biomass ash are: calcium (Ca), potassium (K), silicon (Si), phosphorus (P), magnesium (Mg), as well as sodium (Na), iron (Fe), and aluminum (Al), usually present in the form of oxides (CaO, K_2O, SiO_2, P_2O_5, MgO, Na_2O, Fe_2O_3, and Al_2O_3) [10]. The proportion of these oxides varies by biomass type. Typical wood ash contains 2–5% P_2O_5, 3–7% K_2O, and 30–40% CaO, reflecting the biological roles of calcium and potassium in trees. In contrast, straw ash may contain up to 70% SiO_2 along with significant levels of K_2O, due to high silicon uptake by grasses. Poultry litter ash can exceed 10% P_2O_5, making it a concentrated source of phosphorus. Calcium is largely bound in xylem cell walls and therefore concentrates in wood ash. Potassium is more mobile, residing mainly in bark and foliage; accordingly, its concentration in ash derived from clean stem wood is modest. In contrast, straw and grass ashes are usually rich in silica (SiO_2). In the case of rice ash or rice straw ash, the SiO_2 content can reach 70–80%, as cereals intensively absorb silicon from the soil [1]. Ashes from other types of straw (wheat, rye) also have a high silica content (even > 50%), although usually slightly lower than rice, and significant amounts of potassium. Ashes from animal waste (e.g., from the incineration of chicken manure, dried slurry) are distinguished by a very high concentration of phosphorus; so-called poultry litter ash may contain 5–10% P_2O_5 or more [11]. Similarly, ash from sewage sludge is a valuable source of phosphorus (14–17% P_2O_5) and calcium, making it potentially useful as a fertilizer raw material. The composition of ash can vary not only between different types of biomass, but also within a single type (e.g., wood ash from burning clean wood will be rich in Ca, but ash from wood waste containing sand or construction debris may have an unusually high proportion of silica or other admixtures). The variability in composition also depends on the growing conditions of the plants (mineral content of the soil), the part of the plant (e.g., bark, wood), and even the age of the plants [12–14].

In addition to the main mineral components, biomass ash also contains trace elements and contaminants in trace or moderate amounts. Beneficial micronutrients present in the ash include manganese (Mn), zinc (Zn), copper (Cu), boron (B), and molybdenum (Mo). These trace elements are essential for plant metabolism and add appreciable fertilizer value to the ash. Besides macronutrients, biomass ash supplies beneficial micronutrients—Mn, Zn, Cu, B, and Mo—important for plant metabolism. It may also contain trace heavy metals (Cd, Pb, Cr, Ni, and Hg). Levels vary with feedstock: clean wood ash is typically low in these metals, whereas sewage-sludge or municipal-waste ash can be far higher, so routine monitoring is essential

[1, 12]. For instance, concentrations of mercury in the order of several mg/kg have been found in sewage sludge ash, which raises concerns from an environmental perspective [15]. Ash from wood and agricultural biomass is considered to be fairly "clean" in this respect. Most wood ash meets the heavy metal content standards for fertilizers in many countries. Ash from animal or municipal biomass often exceeds acceptable standards and requires caution before possible agricultural use [1]. In terms of chemical composition, the chlorine and sulfur contents are also important. Biomass usually contains little sulfur (much less than coal), but it can accumulate chlorine, especially straw and plants from chloride soils. Biomass ash can contain significant amounts of chlorine (e.g., in the form of KCl, $CaCl_2$), which promotes its reactivity but also corrosion of installations and may limit certain applications (excess chlorine is harmful to plants and affects the environment).

Because nearly all nitrogen volatilizes during combustion, primarily as inert N_2 with trace NO_x, biomass ash is essentially N-free; therefore, external mineral nitrogen must be supplied when it is used as fertilizer. This constraint prevents ash from serving as a stand-alone fertilizer, yet it remains valuable for providing other macro- and micronutrients [16]. Ash provides a range of other macro- and microelements and acts as a soil deacidifier (thanks to its high CaO and K_2O content, which react with water to form bases). The next section discusses the potential benefits of using biomass ash in agriculture and the challenges associated with it [17].

2.3 Benefits and Challenges of Using Biomass Ash in Agriculture

The use of biomass ash in agriculture is attracting growing interest due to the possibility of recovering and returning valuable nutrients to the soil, while at the same time managing waste in an environmentally friendly manner. The main benefits of such use are presented below, followed by a discussion of the potential risks and limitations (Fig. 2.1).

The main agronomic advantages and technical risks associated with the use of biomass ash are summarized in Table 2.1. This overview can serve as a practical guide for identifying favorable use cases, anticipating site-specific limitations, and informing management strategies.

Biomass ash contains significant amounts of macro- and microelements necessary for plant growth. Particularly important are calcium, potassium, phosphorus, magnesium, as well as trace elements such as copper, zinc, and boron. Adding ash to the soil can compensate for deficiencies in these components. For instance, wood ash is rich in calcium and potassium, which are key elements for plant cell structure and water management. It acts as a natural source of calcium and potassium. Field and laboratory studies confirm that the application of ash (in appropriate doses) increases the content of absorbable forms of potassium, phosphorus, and magnesium in the soil, which translates into a better supply of these elements to plants [1]. For

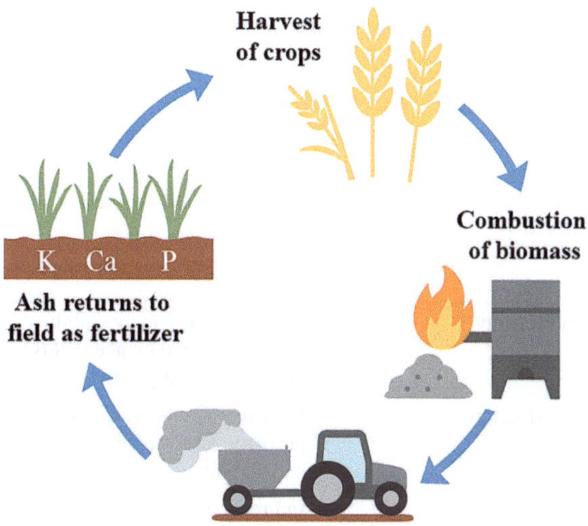

Fig. 2.1 Using biomass ash in agriculture

instance, in experiments with corn and other crops, ash from burning rapeseed straw or grain successfully replaced traditional ammonium phosphate, providing a comparable amount of phosphorus available to plants [11]. The elements absorbed by the plant from the soil end up in the ash after combustion, so adding it to the soil closes the biogenic cycle and reduces soil depletion. The literature emphasizes that this recycling approach is in line with the idea of sustainable agriculture and a circular economy, reducing the need for artificial fertilizers [18, 19].

Most biomass ash is strongly alkaline (pH even above 12 in aqueous solution), which is due to the presence of carbonates, oxides, and hydroxides of calcium and potassium. Ash raises the pH of acidic soils, contributing to their deacidification [1]. Neutralizing soil acidity is important for the release of certain nutrients and overall fertility (in soils with a pH close to neutral, soil microorganisms are more active and many elements are more easily absorbed) [20]. In addition to raising the pH, ash can improve soil structure by introducing calcium and silica, which promote soil particle aggregation. Some studies have shown that adding wood ash to light, sandy soils improved their water capacity and physical properties, which can be attributed to the formation of more durable soil aggregates and calcium supplementation. In addition, ash provides alkaline cations (Ca^{2+}, Mg^{2+}, and K^{+}), which displace acidic ions (H^{+}, Al^{3+}) from the soil sorption complex. The soil becomes more fertile and friendly to root systems. The use of ash can improve yields on acidic and poor soils [2, 21].

The use of ash can also indirectly have a beneficial effect on soil biology. Thanks to deacidification and enrichment with nutrients, the activity of soil microorganisms responsible for the decomposition of organic matter and the release of nutrients for plants increases [11]. Demeyer et al. found that moderate doses of wood ash stimulated mineralization processes in the soil and increased the availability of phosphorus and nitrogen as a result of intensified microbial activity [22]. Well-burned biomass ash contains less than about 2% residual carbon, mostly amorphous and

Table 2.1 Summary of benefits and challenges of using biomass ash in agriculture

Aspect	Benefits	Challenges/risks
Nutrient content	Source of Ca, K, P, Mg and micronutrients (e.g., Zn, B); supports nutrient recycling	Low or variable N content; may require supplementation with nitrogen
Soil pH	Strong alkalinity neutralizes soil acidity	Risk of over-alkalization (especially on neutral/alkaline soils); pH > 8 may reduce micronutrient uptake
Soil fertility	Improves nutrient availability and CEC; can enhance yields on acidic and degraded soils	Effect may be short-term; reapplication may be needed after several years
Soil structure	Improves aggregation and porosity; adds Ca and silica for soil stability	High Na content may cause soil dispersion and salinization (site-specific)
Soil biology	Indirect stimulation of microbial activity through pH correction and nutrient supply	Temporary disruption to sensitive organisms (e.g., earthworms, fungi)
Climate mitigation	Potential for CO_2 sequestration via carbonation reactions (formation of $CaCO_3$, $MgCO_3$)	Carbon sequestration potential is limited compared to biochar; depends on ash reactivity and conditions
Waste valorization	Converts biomass combustion waste into a valuable product; reduces landfill and disposal costs	Legal classification as "waste" complicates use; requires regulatory alignment
Variability in composition	Some ashes rich in P or K can replace mineral fertilizers selectively	Strong batch-to-batch variability; requires testing, labeling and quality control
Contaminant content	Ash from clean biomass often meets safety standards	Risk of heavy metals (Cd, Pb, Hg), dioxins in contaminated ash; requires screening and potential pre-treatment
Fertilizer integration	Can be blended with compost, biochar, or organics to create custom fertilizers	Difficult to standardize formulations without consistent ash composition

readily weathered, so it does not replicate the long-term soil-carbon benefits associated with engineered biochar. There are hypotheses that mineral carbon from ash may contribute to a permanent increase in organic carbon content in the soil through CO_2 sequestration mechanisms. The biomass ash has the ability to bind CO_2 from the atmosphere after application to the soil. As a result of the carbonation of free CaO and MgO, carbon dioxide is absorbed and carbonates ($CaCO_3$, $MgCO_3$) are formed. Laboratory carbonation of Ca-rich wood ash commonly captures 60–80 g CO_2 per kg under ambient conditions, with higher uptakes under controlled humidity [23]. Therefore, the use of ash in agriculture may have an additional positive effect in the form of a small contribution to carbon sequestration and climate change mitigation (although this is an additional aspect, depending on environmental conditions).

Another undoubted benefit is the fact that the agricultural use of ash allows waste products to be recycled, reducing the need for storage. Biomass ash is often treated as waste under the law, which generates disposal costs. Using it as fertilizer or soil improver transforms waste into a value-added product, while saving landfill space and reducing the risk of leachate contamination. In addition, partially replacing mineral fertilizers with ash can bring economic benefits to farms (a cheaper source of nutrients than synthetic fertilizers) and conserve natural resources, particularly valuable in the case of phosphorus, whose global reserves are limited [11]. Every ton of P-rich ash used on fields means less phosphate rock used in industrial fertilizers. From the perspective of biomass energy, the possibility of returning ash to farmers can also improve the ecological balance of bioenergy (full utilization of the process product) and build a symbiosis between the energy sector and agriculture.

Biomass ash has a number of beneficial properties from an agricultural point of view: it provides calcium, potassium, phosphorus, and microelements, deacidifies the soil, improves its properties, closes the nutrient cycle, and helps manage waste. Scientific experiments confirm that when used properly, it can increase yields and improve soil health [2]. For ash use to be safe and effective, the potential challenges and risks described below must also be taken into account.

The most serious concerns related to the agricultural use of ash concern the presence of harmful elements, especially heavy metals (Cd, Pb, Hg, Cr, etc.) and possibly toxic compounds (e.g., dioxins, if the biomass burned was contaminated). The accumulation of heavy metals in soils and plants poses a threat to the ecosystem and human health, which is why many countries have strict regulations limiting their permissible concentrations in fertilizers and soils [7]. As mentioned, ash from pure forest biomass usually meets the standards, but ash from lower quality biomass (sludge, agricultural waste with admixtures) may exceed these limits [24]. If such ash is spread on fields, heavy metals can enter the food chain. First, they accumulate in the soil, then they are absorbed by plants and end up in food or feed [25]. For instance, excessive cadmium content in soil promotes its accumulation in cereal grains, which has negative health effects. Therefore, it is necessary to control the quality of ash intended for agriculture and to use only ash that meets safety standards [1]. In practice, this means testing each batch of ash for heavy metal content (and possibly other contaminants, such as excess chlorides or sulfates) and only allowing safe material to be used. Another reliable solution to minimize the problem of metals is the selection of ash fractions. As discussed, metals are mainly concentrated in fine fly ash, so by using mainly bottom ash (coarser), the introduction of metals can be reduced [7]. If the ash does not meet the requirements (e.g., too high levels of Pb, Cd), it should be treated as waste to be disposed of or, alternatively, subjected to purification processes (e.g., washing, extraction) before it is released into the environment [1].

Biomass ash contains significant amounts of salts, primarily compounds of potassium, calcium, and sodium which, if over applied, can lead to soil salinization and osmotic stress in plants. High sodium levels, in particular, especially from biomass grown in saline environments or exposed to seawater can degrade soil structure and fertility. Therefore, site-specific dosage and careful monitoring are essential to prevent adverse effects [26]. Ash used in excessive doses can also rapidly raise soil pH above the optimal range, leading to over-alkalization. Crops generally prefer a pH range of 6–7.5; much higher values (pH 8–9) can limit the availability of micronutrients (e.g., iron, manganese) [27]. In addition, soluble ash components (especially potassium) are easily washed away by rainwater. If ash is mixed with the soil before planting, a significant portion of the potassium may be quickly washed into the soil profile or groundwater before the plants have a chance to absorb it [2]. This results in the loss of valuable nutrients (reduced fertilizing effect) and potential water pollution (excess K and other ions). For this reason, it is important to select the appropriate dose of ash depending on the type of soil and fertilization needs, as well as to avoid application before heavy rainfall. As a rule, doses of several tons of ash per hectare every few years are recommended, rather than very high single doses. The literature also emphasizes the need to monitor soil parameters (pH, conductivity as a measure of salinity) after applying ash to adjust further fertilization if necessary [1].

Another challenge is the heterogeneity of ashes. Their composition can vary significantly from batch to batch, making it difficult to standardize dosages and predict effects. A farmer applying manure knows exactly the NPK content of the product. In the case of biomass ash, the composition may be variable and not fully known without laboratory analysis. It may turn out that one shipment of wood ash has 5% K_2O and 2% P_2O_5, while another (from another plant) has 2% K_2O and trace P, respectively, depending on the species of wood burned or bark admixtures, for example. The lack of a uniform composition causes difficulties in dosing for specific fertilizer needs. The solution is to analyze and label the ash each time for the content of the main fertilizer components, to be able to calculate the supplemental dose of fertilizers (especially nitrogen, as mentioned). In the European Union, work is underway to include ash in fertilizer regulations—the new EU Fertilizer Products Regulation (2019/1009) allows, among other things, biomass ash as a fertilizer ingredient, provided that purity criteria and composition declarations are met [28]. The diversity of ashes makes it necessary to develop a classification. A proposal for such a categorization of ashes according to their main nutrients was presented, for example, by IEA Bioenergy experts. They distinguished between C-type (rich in Ca—calcium), K-type (rich in K—potassium), P-type (phosphorus), and Si-type (silica) ashes [29]. This division can help determine fertilizer potential, e.g., K-type ash is suitable as a potassium fertilizer, but will not provide phosphorus; P-type ash (e.g., from poultry, sludge) can replace phosphate, but has little potassium, etc. Variability in composition is also a problem when mixing ash with other fertilizers or wastes (e.g., manufacturing granular blended fertilizers). Each batch may require different proportions, which complicates the technological process.

Despite its many advantages, the addition of ash can cause some negative side effects. A sharp change in soil chemistry (pH, salinity) affects soil organisms. Some sensitive species of soil microorganisms or mesofauna (e.g., earthworms) may be temporarily displaced or reduced in activity immediately after the ash application, before recolonization occurs at the new pH. The high alkalinity of ash can temporarily reduce the availability of certain micronutrients (Fe, Mn, Zn) to plants, manifesting as leaf chlorosis until conditions stabilize [2]. Another aspect is the leaching of substances from the ash: in addition to the desired K or Ca, the ash contains, for example, sulfates, chlorine, and sodium. Excess of these anions and cations can run off into waterways, causing mineralization. This is especially true on light soils and with high rainfall. In the long term, there is a question about the permanence of the ash effect, with studies indicating that the increase in pH and assimilates after a single treatment can disappear after several years, necessitating repeat applications [1].

Biomass ash has high potential as a liming and fertilizing agent in agriculture. It allows nutrient recycling and improves soil properties [11, 30]. At the same time, it requires a conscious and careful approach due to the variability of composition and the presence of potentially harmful admixtures. The key to success is quality assurance (compositional testing, elimination of contaminated ash) and adjustment of dosages to soil needs and conditions. Current research and implementation projects focus on developing mixtures of ash with other materials (e.g., biochar, compost), granulation of ash to facilitate application, and monitoring the long-term effects of application on the soil ecosystem [1]. All of this is aimed at maximizing benefits (soil fertility, closed cycle of elements, and reduced use of fertilizers) while minimizing risks (accumulation of metals, chemical imbalances).

References

1. I. Maj, K. Niesporek, P. Płaza, J. Maier, P. Łój, Biomass ash: a review of chemical compositions and management trends. Sustainability **17**(11), 4925 (2025). https://doi.org/10.3390/su17114925
2. Z. Romanowska-Duda, R. Janas, M. Grzesik, Increasing fertilization efficiency of biomass ash by the synergistically acting digestate and extract from water plants sequestering CO_2 in sorghum crops. Molecules **29**(18) (2024). https://doi.org/10.3390/molecules29184397
3. I. Barišić, I.N. Grubeša, D.K. Hackenberger, G. Palijan, S. Glavić, M. Trkmić, Multidisciplinary approach to agricultural biomass ash usage for earthworks in road construction. Materials **15**(13) (2022). https://doi.org/10.3390/ma15134529
4. L. Tosti, A. van Zomeren, J.R. Pels, R.N.J. Comans, Evaluating biomass ash properties as influenced by feedstock and thermal conversion technology towards cement clinker production with a lower carbon footprint. Waste Biomass Valorization **12**(8), 4703–4719 (2021). https://doi.org/10.1007/s12649-020-01339-0
5. I. Maj, S. Kalisz, S. Ciukaj, Properties of animal-origin ash—a valuable material for circular economy. Energies (Basel) **15**(4) (2022). https://doi.org/10.3390/en15041274
6. I. Maj, Significance and challenges of poultry litter and cattle manure as sustainable fuels: a review. Energies (2022). https://doi.org/10.3390/en15238981
7. C. Lanzerstorfer, Grate-fired biomass combustion plants using forest residues as fuel: enrichment factors for components in the fly ash. Waste Biomass Valorization **8**(1), 235–240 (2017). https://doi.org/10.1007/s12649-016-9565-6
8. C. Lanzerstorfer, Chemical composition and physical properties of filter fly ashes from eight grate-fired biomass combustion plants. J. Environ. Sci. **30**, 191–197 (2015). https://doi.org/10.1016/J.JES.2014.08.021
9. H. Nurmesniemi, M. Mäkelä, R. Pöykiö, K. Manskinen, O. Dahl, Comparison of the forest fertilizer properties of ash fractions from two power plants of pulp and paper mills incinerating biomass-based fuels. Fuel Process. Technol. **104**, 1–6 (2012). https://doi.org/10.1016/J.FUPROC.2012.06.012

10. A. Fuller, M. Carbo, P. Savat, J. Kalivodova, J. Maier, G. Scheffknecht, Results of fly ash quality for disposal options from high thermal shares up to pure biomass combustion in a pilot-scale and large scale pulverized fuel power plants. Renew. Energy **75**, 899–910 (2015). https://doi.org/10.1016/J.RENENE.2014.10.032
11. K. Schiemenz, B. Eichler-Löbermann, Biomass ashes and their phosphorus fertilizing effect on different crops. Nutr. Cycl. Agroecosyst. **87**(3), 471–482 (2010). https://doi.org/10.1007/S10705-010-9353-9/TABLES/9
12. M. Coutand, M. Cyr, P. Clastres, Use of sewage sludge ash as mineral admixture in mortars. Proc. Inst. Civ. Eng. Constr. Mater **159**(4), 153–162 (2006). https://doi.org/10.1680/COMA.2006.159.4.153
13. P. Sakiewicz et al., Innovative technological approach for the cyclic nutrients adsorption by post-digestion sewage sludge-based ash co-formed with some nanostructural additives under a circular economy framework. Int. J. Environ. Res. Public Health **19**(17), 11119 (2022). https://doi.org/10.3390/IJERPH191711119
14. M. Kasina, K. Jarosz, M. Stolarczyk, J. Göttlicher, R. Steininger, M. Michalik, Characteristic of phosphorus rich compounds in the incinerated sewage sludge ashes: a case for sustainable waste management. Sci. Rep. **13**(1), 1–11 (2023). https://doi.org/10.1038/S41598-023-36407-7
15. J. Werther, T. Ogada, Sewage sludge combustion. Prog. Energy Combust. Sci. **25**(1), 55–116 (1999). https://doi.org/10.1016/S0360-1285(98)00020-3
16. R.P. Girón, B. Ruiz, E. Fuente, R.R. Gil, I. Suárez-Ruiz, Properties of fly ash from forest biomass combustion. Fuel **114**, 71–77 (2013). https://doi.org/10.1016/J.FUEL.2012.04.042
17. H. Górecka, K. Chojnacka, H. Górecki, The application of ICP-MS and ICP-OES in determination of micronutrients in wood ashes used as soil conditioners. Talanta **70**(5), 950–956 (2006). https://doi.org/10.1016/J.TALANTA.2006.05.061
18. G. Zając, J. Szyszlak-Bargłowicz, W. Gołębiowski, M. Szczepanik, Chemical characteristics of biomass ashes. Energies **11**(11), 2885 (2018). https://doi.org/10.3390/EN11112885
19. Z. Romanowska-Duda, R. Janas, M. Grzesik, B. van Duijn, Valorization of sorghum ash with digestate and biopreparations in the development biomass of plants in a closed production system of energy. Sci. Rep. **13**(1), 1–16 (2023). https://doi.org/10.1038/S41598-023-45733-9
20. *Best Management Practices for Wood Ash as Agricultural Soil Amendment | UGA Cooperative Extension.* Accessed 12 June 2025. [Online]. Available: https://extension.uga.edu/publications/detail.html?number=B1142&title=best-management-practices-for-wood-ash-as-agricultural-soil-amendment
21. N.L. Ukwattage, P.G. Ranjith, M. Bouazza, The use of coal combustion fly ash as a soil amendment in agricultural lands (with comments on its potential to improve food security and sequester carbon). Fuel **109**, 400–408 (2013). https://doi.org/10.1016/J.FUEL.2013.02.016
22. A. Demeyer, J.C. Voundi Nkana, M.G. Verloo, Characteristics of wood ash and influence on soil properties and nutrient uptake: an overview. Bioresour. Technol. **77**(3), 287–295 (2001). https://doi.org/10.1016/S0960-8524(00)00043-2
23. C. Wang et al., Accelerated CO_2 mineralization technology using fly ash as raw material: recent research advances. Chem. Eng. J. **488**, 150676 (2024). https://doi.org/10.1016/J.CEJ.2024.150676
24. A.S.M. Aljohani, Heavy metal toxicity in poultry: a comprehensive review. Front. Vet. Sci. **10**, 1161354 (2023). https://doi.org/10.3389/FVETS.2023.1161354/XML/NLM
25. Y. Chai et al., Valorization of heavy metal contaminated biomass: recycling and expanding to functional materials. J. Clean. Prod. **366**, 132771 (2022). https://doi.org/10.1016/J.JCLEPRO.2022.132771
26. E. Boros-Lajszner, J. Wyszkowska, J. Kucharski, Effect of ash from *Salix viminalis* on the biomass and heating value of *Zea mays* and on the biochemical and physicochemical properties of soils. Energies **16**(24), 8037 (2023). https://doi.org/10.3390/EN16248037
27. J.I. Odzijewicz, E. Wołejko, U. Wydro, M. Wasil, A. Jabłońska-Trypuć, Utilization of ashes from biomass combustion. Energies **15**(24), 9653 (2022). https://doi.org/10.3390/EN15249653
28. EP, *Regulation (EU) 2019/1009 of the European Parliament and of the Council of 5 June 2019 Laying Down Rules on the Making Available on the Market of EU Fertilising Products and*

Amending Regulations (EC) No 1069/2009 and (EC) No 1107/2009 and Repealing Regulation (EC) No 2003/2003 (2019)
29. M. Cremers (Dnvgl), D. Matschegg, Birgitte Primdal Dam (Emineral A/S), Roberta Roberto (ENEA), Rob Mager (Ontario Power Generation), Kent Davidsson (2020)
30. M. Mozaffari, M.P. Russelle, C.J. Rosen, E.A. Nater, Nutrient supply and neutralizing value of alfalfa stem gasification ash. Soil Sci. Soc. Am. J. **66**(1), 171–178 (2002). https://doi.org/10.2136/SSSAJ2002.1710

Chapter 3
Biochars

3.1 Introduction

Biochar is a solid, porous carbon material obtained from biomass through pyrolysis, a thermochemical process carried out under conditions of limited oxygen access [1, 2]. In practice, biochar is a type of charcoal of plant origin, produced from various organic raw materials, such as wood waste, straw and crop residues, animal manure, or sewage sludge [3]. During pyrolysis, organic material is subjected to high temperatures (usually 300–700 °C) in an oxygen-free atmosphere, resulting in its decomposition into three fractions: gaseous (syngas), liquid (bio-oil), and solid (biochar) [2]. The proportions of these products depend on the process conditions (Fig. 3.1). Slow pyrolysis (at lower heating temperatures and longer residence times) promotes the formation of larger amounts of biochar at the expense of lower bio-oil and gas production. On the other hand, fast pyrolysis (with rapid heating of biomass) results in a greater amount of liquid products, while gasification of biomass at high temperatures and in the presence of a small amount of oxidizer leads mainly to the production of fuel gas at the expense of the solid fraction [4]. The literature also distinguishes between different types of pyrolysis processes, such as torrefaction (mild pyrolysis at a lower temperature) and hydrothermal conversion (HTC, wet pyrolysis for materials with high moisture content), which also allow carbonaceous material with biochar properties to be obtained [1, 5]. However, biochar is most often obtained by classical pyrolysis in reactors with an inert atmosphere (e.g., in a nitrogen shield) [6].

The parameters of the biochar production process strongly influence the properties of the resulting product. For example, increasing the pyrolysis temperature generally reduces the yield of biochar obtained. Above approx. 500 °C, intensive destruction of lignin and volatilization of volatile components occur, resulting in a significant decrease in the proportion of the solid fraction [1]. At lower temperatures (300–500 °C), higher biochar yields are obtained and more biogenic elements (nitrogen, sulfur, potassium, etc.) are retained, which at higher temperatures are

K. Chojnacka et al., *Sustainable Fertilizers: Utilizing Biomass Ashes and Biochars*, SpringerBriefs in Applied Sciences and Technology,
https://doi.org/10.1007/978-3-032-03765-7_3

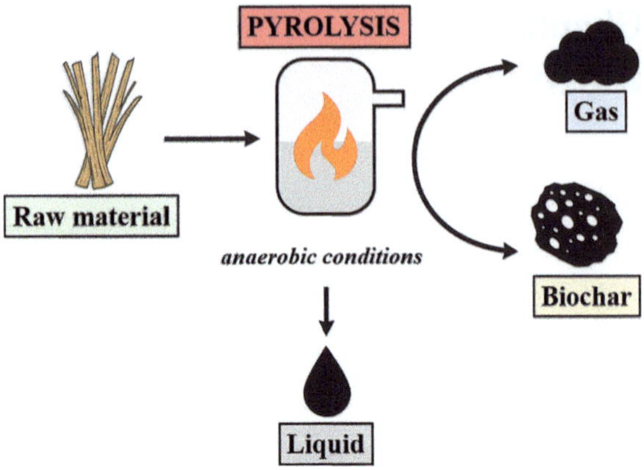

Fig. 3.1 Biomass pyrolysis process scheme

largely degassed or burned [7, 8]. On the other hand, longer residence time of the material in the reactor and lower heating rate (characteristic of slow pyrolysis) favor the production of biochar with a more developed porous structure and greater carbon stability, although the process itself is less time-efficient. Different types of biomass as raw material may require different process conditions, e.g., material with high moisture content may first be dried or processed using a hydrothermal method, while lignocellulosic biomass (wood, straw) works well in classic pyrolysis. In summary, biochar is defined as a product of biomass pyrolysis under oxygen-deficient conditions, whereby control of process parameters (temperature, time, and heating rate) and selection of raw material allow for the production of material with specific characteristics and efficiency [2, 7].

3.2 Chemical Composition and Properties of Biochars

Biochar has unique physicochemical properties that distinguish it from both its source material (biomass) and typical organic matter in soil. The elemental composition of biochar is dominated by carbon, which usually accounts for approximately 50% to as much as 80–90% of the mass of biochar [9]. The elemental composition of biochar generally includes carbon, hydrogen, nitrogen, sulfur, and oxygen, as well as inorganic minerals such as potassium, phosphorus, calcium, magnesium, iron, silicon, and sodium, which are present in the ash fraction. The ash also contains mineral components derived from biomass, including potassium, phosphorus, calcium, magnesium, silicon, sodium, and iron [10]. Their concentration in biochar is usually higher than in the raw material due to the loss of organic matter during pyrolysis. The thermal process leaves non-degradable mineral components in the form of ash. The nature of carbon in biochar differs from that in typical biomass: aromatic structures (condensed carbon rings) dominate, giving biochar high chemical stability and resistance to further decomposition [11]. Pyrolysis converts aliphatic and cellulosic carbon into a much more stable aromatic structure that resembles fossil carbon (amorphous and partially crystalline form of aromatic carbon). This makes biochar a durable material in the soil environment [12].

Physical Properties. Biochar particles are light, black granules with hierarchical porosity (nano → macro); BET surface areas often exceed 200 $m^2 g^{-1}$, underpinning superior water and nutrient retention [3].

Chemical Properties. Wood-derived biochars are typically alkaline (pH 7–10) owing to carbonate/oxide ash; manure or straw biochars add significant P and K but raise electrical conductivity [11, 12].

Functional Properties. Cation-exchange capacity spans 5–150 cmol kg^{-1} and declines as H/C_org falls below 0.4. These traits dictate whether a product is marketed as a *fertilizer* (nutrient-rich) or as a *carrier/conditioner* (nutrient-poor yet highly sorptive) [13].

Another important feature of biochar is the abundance of functional groups on its surface. Although a significant proportion of heteroatoms volatilize during pyrolysis, certain reactive functional groups remain on the surface of biochar. This is particularly true for biochar obtained at lower temperatures. These include, among others, acidic carboxyl, phenolic, and carbonyl groups, as well as basic groups, e.g., pyrones [14]. The presence of these chemical groups gives the surface of biochar an electrical charge (usually predominantly negative at neutral and alkaline pH) and determines its exchange capacity. In practice, biochar often has a significant cation exchange capacity (CEC). It can adsorb and exchange nutrient cations such as K^+, Ca^{2+}, Mg^{2+}, and NH_4^+ ammonium on the surface of its particles [10, 15]. High CEC is particularly typical for biochar produced at lower temperatures (below ~ 500 °C), as it retains more surface oxygen in the carboxyl and phenolic groups responsible for the negative charge. As the pyrolysis temperature increases, the number of acidic groups decreases (they undergo thermal removal), which means that very high-temperature biochar (> 700 °C) may have a lower CEC and even show a predominance of positive charges on the surface (anion exchange capacity). Nevertheless, most typical biochar contributes a beneficial sorption center to the soil, binding nutrient cations and preventing their leaching [10].

Another important property of biochar is its pH. Most biochars have a neutral or alkaline pH [16]. The alkaline reaction stems from basic ash constituents-carbonates and oxides of alkali and alkaline-earth metals-and from the elimination of acidic functional groups during pyrolysis [17]. Adding biochar to acidic soil usually raises the soil pH (similar to liming), which can be beneficial for plant growth in overly acidic soils. Biochar produced from certain raw materials or specially chemically activated biochar can be acidic. There are studies where the addition of such acidic biochar slightly lowered the pH of calcareous soil. Biochar is considered a deacidifying and pH-buffering component in soil systems [10, 18, 19].

The chemical composition of biochar depends largely on the feedstock. Biochar from ash-rich biomass (e.g., straw, rice husks, animal manure, or algae) will contain more macronutrients such as phosphorus, potassium, calcium, and magnesium, while wood biochar from pure wood is poor in these elements and consists mainly of carbon [13]. For instance, biochar from animal waste and manure has a higher nitrogen and phosphorus content than wood biochar [20, 21]. At the same time, higher pyrolysis temperatures promote the loss of nitrogen and sulfur—these elements escape in gaseous form, which is why high-temperature biochar has very low N levels (often < 0.5%) [22]. In turn, the mineral ash content (i.e. P, K, Ca, Mg, etc.) increases with temperature, as more organic matter is burned, leaving behind inorganic residue [23–25]. Process optimization can therefore allow the desired components to be retained. For instance, by conducting pyrolysis at temperatures below 400 °C, a significant proportion of nutrients (N, K, and S) can be retained in biochar, whereas higher temperatures remove them almost completely [7].

Biochar supplies beneficial micronutrients (Mn, Zn, Cu, B, Mo) but can also carry trace metals (Cd, Pb, Cr, Ni, and Hg); levels are low in clean-wood char and higher in sludge-derived char, so every batch must satisfy EBC and EU CMC 14 limits [3]. Both frameworks cap the sum of the 16 EPA PAHs at < 6 mg kg^{-1} DM (EU) or < 12 mg kg^{-1} DM (EBC Class B). These targets are typically met when production temperatures are ≥ 500 °C and the char is rapidly quenched to minimize soot condensation.

Biochar has the ability to sorb pollutants. Thanks to its porosity and functional groups, it can bind both heavy metals and organic toxins (e.g., pesticides, PAHs) inside itself. This is used in remediation. The addition of biochar to contaminated soils immobilizes metals and pollutants, reducing their bioavailability [18, 26]. Ash-rich biochar (e.g., from agricultural waste) often contains carbonate forms of calcium and potassium, which increase soil alkalinity and electrical conductivity, which in excess can adversely affect sensitive plants. Therefore, for agricultural purposes, biochar from relatively clean biomass (wood, straw) is preferred, possibly rinsed or seasoned to wash out excess salts [10].

3.3 Benefits of Biochars as a Fertilizer

Fertilizer Versus Carrier. Unmodified wood biochar is nutrient-poor and functions chiefly as a conditioner. Manure- or straw-derived biochars, or chars pre-loaded with mineral solutions, act as slow-release fertilizers. Clear labeling of the intended role is essential for correct agronomic dosing [27]. Depending on its source and preparation, biochar may function either as a nutrient-poor conditioner or as a slow-release fertilizer. Table 3.1 summarizes the agronomic functions of different types of biochar and their implications for nutrient management strategies.

The main agronomic benefits of using biochar include: improving the physical properties of the soil (structure, water capacity, aeration), improving chemical properties (increasing pH, increasing sorption capacity, retaining nutrients), and stimulating soil biological activity (development of soil microorganisms). Biochar used as a soil additive can increase crop yields, especially on poor and degraded soils [7, 18].

Improving fertility and retaining nutrients. Biochar acts as a store and carrier of nutrients in the soil. Thanks to its high cation exchange capacity, it binds ions of elements such as potassium, magnesium, calcium, and ammonium, preventing

Table 3.1 Agronomic roles of biochar depending on source and treatment

Type of biochar	Primary function	Nutrient content	Agronomic use
Unmodified wood biochar	Soil conditioner	Very low (nutrient-poor)	Improves structure, water retention, pH
Manure-derived biochar	Slow-release fertilizer	Moderate to high	Nutrient source and conditioner
Straw-derived biochar	Slow-release fertilizer	Moderate	Useful on nutrient-poor soils
Biochar pre-loaded with mineral NPK	Enhanced nutrient carrier	High (depends on loading)	Acts like slow-release fertilizer
Biochar co-composted with organics	Integrated fertilizing agent	Moderate to high	Combines stability of biochar with nutrient availability of compost

them from being washed deep into the soil profile [22]. Research indicates that the addition of biochar significantly reduces nitrogen losses from the soil. It limits both the leaching of nitrates into groundwater and the emission of ammonia and nitrous oxide (greenhouse gases) into the atmosphere. For instance, in the presence of biochar, nitrogen losses from fertilizers through volatilization and leaching are significantly lower, which translates into higher efficiency of nitrogen fertilizer use by plants [13]. Biochar releases certain bound components slowly, acting similarly to slow-release fertilizers. Fertilizers enriched with biochar (so-called biochar-based fertilizers) are considered a promising solution. Through sorption of mineral fertilizer nutrients onto biochar, it is possible to obtain a form that releases nutrients slowly and is less prone to loss. For example, biochar loaded with nitrogen, phosphorus, and potassium compounds during composting can act as an effective long-acting fertilizer [22, 27]. In addition, the high porosity of biochar promotes water and nutrient retention in the root zone. Sandy soils with biochar additives show less nitrogen and phosphorus leaching into deeper layers. The presence of biochar in the substrate increases water and fertilizer retention, improving the availability of these resources to plants over a longer period of time [28].

The incorporation of biochar into the soil favorably changes its physical properties, which translates into better root growth conditions. Biochar improves soil structure. Its particles, acting similarly to soil organic matter, promote the aggregation of soil mineral particles and prevent excessive soil compaction. On heavy soils, the addition of biochar reduces the bulk density and compactness of the soil, making it more permeable and looser [7, 18, 19]. At the same time, in light (sandy) soils, biochar increases water capacity. Its micropores act like a sponge that retains water, which delays the drying of the soil profile [29]. Water infiltration is also improved. Biochar increases rainwater infiltration, reducing surface runoff and soil erosion. Due to modification of soil porosity and structure, biochar provides better conditions for root system development (roots penetrate looser soil with better aeration) and increases the availability of both water and oxygen in the root zone. All this translates into a healthier, more developed root system and, a better supply of water and nutrients to plants [18, 19].

Impact on microorganisms and soil health. One of the less direct but very important effects of using biochar is the stimulation of beneficial soil microflora. The porous structure of biochar creates microhabitats for microorganisms. Bacteria and fungi can settle in the pores of biochar, finding shelter from soil predators and access to nutrients sorbed on biochar [3, 12]. Numerous studies confirm that the addition of biochar increases soil microbial activity and promotes beneficial microorganisms. For instance, a significant increase in the number of bacteria has been observed in soils with biochar (a meta-analysis showed an increase in soil bacterial diversity, with no negative impact on fungi) [30]. Biochar can also stimulate symbiotic microorganisms. Experimental studies have reported enhanced root colonization by mycorrhizal fungi and higher numbers of beneficial rhizosphere bacteria after biochar application [3]. The resulting microbial community improves plant nutrition through atmospheric nitrogen fixation and the solubilization of otherwise unavailable phosphorus compounds, while also shielding roots from pathogens. Biochar can contain trace pyrolysis-derived compounds that stimulate microbial and plant growth; together with the porous carbon matrix this often raises soil enzymatic activity, faster nutrient cycling and more favorable conditions for plant development [12, 13].

The consequence of the above-mentioned effects is a noticeable improvement in plant growth and increased yields on soils enriched with biochar. Many field and pot studies have shown that biochar promotes plant growth both indirectly (by improving soil conditions) and directly (by providing certain nutrients, such as potassium and calcium). Literature reviews indicate that the typical increase in yield after the application of biochar averages several percent, with the strongest effects on soils that are initially poor—acidic, low in organic matter and nutrients [18]. A meta-analysis of studies showed an average global increase in yields of ~ 14% after the application of biochar, with an increase of approximately 12% in the short term ($\leq$ 5 years) and an average increase of 21% in long-term experiments (> 5 years) [31]. This is confirmed by another meta-analysis, which showed that biochar significantly increases yields in the tropics, while in temperate climates, on average, no statistically significant improvement in yields was observed. This is because soils in the tropics are more often acidic and degraded, so they benefit more from the addition of biochar, while in temperate regions, arable soils are often richer and better regulated, so the effect of biochar may be neutral. Nevertheless, in specific situations, benefits have also been observed in temperate regions, e.g., on sandy soils with low sorption capacity, biochar significantly improved water and fertilizer retention, translating into better plant supply and higher grain yields [31].

Biochar does not usually replace traditional fertilizers entirely, but works in tandem with other fertilization methods. The best results are achieved when biochar is combined with mineral fertilizers or organic matter. For instance, a meta-analysis showed that the simultaneous use of biochar and fertilizers (mineral or organic) can increase yields more than either intervention alone. In one field study, the combination of biochar and nitrogen fertilizer resulted in a 179% increase in yield compared to the control without any additives [32]. Biochar acts as a buffer here, retaining fertilizer components and releasing them gradually, increasing their efficiency of use by plants [22]. Co-composting biochar with organic waste also yields excellent results: adding biochar to compost accelerates the composting process and reduces nitrogen (ammonia) losses during fermentation, and the resulting carbon-enriched compost has a higher fertilizing value. Such a product combines the advantages of biochar and compost. It provides organic matter and microelements, while also having the high stability and sorption capacity of biochar. Research shows that compost with added biochar leads to better plant growth than compost or biochar alone. This is because compost provides readily available nutrients in the early stages of cultivation, while biochar takes on a long-term role. It gradually releases adsorbed substances and maintains fertility in subsequent phases [7].

Additional environmental benefits. In addition to its direct impact on soil fertility and crop yields, the use of biochar offers a number of environmental benefits. One of the most important is carbon sequestration in the soil. The introduction of stable, carbon-based biochar into the soil profile results in the long-term accumulation of organic carbon, which, in the form of CO_2, will not be released into the atmosphere [12]. It is estimated that the widespread use of biochar in agriculture could significantly reduce CO_2 concentrations in the atmosphere on a global scale (by storing part of the carbon from biomass in soils) [18]. Biochar contributes to the reduction of nitrous oxide (N_2O) emissions, a powerful greenhouse gas produced in agricultural soils. Meta-analyses confirm that biochar can reduce N_2O emissions from fertilized soils by up to several dozen percent by changing the conditions of denitrification and ammonium nitrogen sorption [33]. In this way, biochar fits into climate-neutral agricultural strategies. At the same time, it improves crop yields and mitigates climate change by absorbing carbon dioxide and reducing greenhouse gas emissions [18].

Biochar as a fertilizer additive offers many benefits: it improves soil properties (chemical, physical, and biological), increases the efficiency of traditional fertilizers, and significantly promotes plant growth and yield, especially in problematic soils. The introduction of biochar into agricultural practice can therefore contribute both to increased crop production and to environmental sustainability. For these reasons, biochar is the subject of growing interest in science and agricultural practice as an innovative component of fertilization systems in the twenty-first century.

References

1. C. Youngsang, K.L. Soo, Y.Y. Hah, W.K. Seung, BioRes_16_3_6512_Chun_LYK_Review_Recent_Biochar_Prodn_Biomass_Classif_Pyrol_18707 (2021)
2. M. Varkolu et al., Recent advances in biochar production, characterization, and environmental applications. Catalysts **15**(3), 243 (2025). https://doi.org/10.3390/CATAL15030243
3. A. Tomczyk, Z. Sokołowska, P. Boguta, Biochar physicochemical properties: pyrolysis temperature and feedstock kind effects. Rev. Environ. Sci. Biotechnol. **19**(1), 191–215 (2020). https://doi.org/10.1007/S11157-020-09523-3/TABLES/3

4. P. Roy, G. Dias, Prospects for pyrolysis technologies in the bioenergy sector: a review. Renew. Sustain. Energy Rev. **77**, 59–69 (2017). https://doi.org/10.1016/J.RSER.2017.03.136
5. D. Pandey, A. Daverey, K. Arunachalam, Biochar: production, properties and emerging role as a support for enzyme immobilization. J. Clean. Prod. **255**, 120267 (2020). https://doi.org/10.1016/J.JCLEPRO.2020.120267
6. K. Weber, P. Quicker, Properties of biochar. Fuel **217**, 240–261 (2018). https://doi.org/10.1016/J.FUEL.2017.12.054
7. T. Allohverdi, A.K. Mohanty, P. Roy, M. Misra, A review on current status of biochar uses in agriculture. Molecules **26**(18), 5584 (2021). https://doi.org/10.3390/MOLECULES26185584
8. S.D. Joseph et al., An investigation into the reactions of biochar in soil. Soil Res. **48**(7), 501–515 (2010). https://doi.org/10.1071/SR10009
9. S. Li, C.Y. Chan, Will biochar suppress or stimulate greenhouse gas emissions in agricultural fields? Unveiling the dice game through data syntheses. Soil Syst. **6**(4), 73 (2022). https://doi.org/10.3390/SOILSYSTEMS6040073
10. S. Fawzy, A.I. Osman, H. Yang, J. Doran, D.W. Rooney, Industrial biochar systems for atmospheric carbon removal: a review (2021). https://doi.org/10.1007/s10311-021-01210-1
11. N.A. Qambrani, M.M. Rahman, S. Won, S. Shim, C. Ra, Biochar properties and eco-friendly applications for climate change mitigation, waste management, and wastewater treatment: a review. Renew. Sustain. Energy Rev. **79**, 255–273 (2017). https://doi.org/10.1016/J.RSER.2017.05.057
12. S. Li, D. Tasnady, Biochar for soil carbon sequestration: current knowledge, mechanisms, and future perspectives. C **9**(3), 67 (2023). https://doi.org/10.3390/C9030067
13. H.M. Alkharabsheh et al., Biochar and its broad impacts in soil quality and fertility, nutrient leaching and crop productivity: a review. Agronomy **11**(5), 993 (2021). https://doi.org/10.3390/AGRONOMY11050993
14. L. Usevičiūtė, E. Baltrėnaitė, Methods for determining lignocellulosic biochar wettability. Waste Biomass Valorization **11**(8), 4457–4468 (2020). https://doi.org/10.1007/S12649-019-00713-X/FIGURES/7
15. S. Mia, B. Singh, F.A. Dijkstra, Aged biochar affects gross nitrogen mineralization and recovery: a ^{15}N study in two contrasting soils. GCB Bioenergy **9**(7), 1196–1206 (2017). https://doi.org/10.1111/GCBB.12430
16. W. Chen, J. Meng, X. Han, Y. Lan, W. Zhang, Past, present, and future of biochar. Biochar **1**(1), 75–87 (2019). https://doi.org/10.1007/S42773-019-00008-3
17. B. Singh, Q. Shen, M.C. Arbestain, M.M. Dolk, M. Camps-Arbestain, Biochar pH, electrical conductivity and liming potential biochar ageing effects on N and P dynamics: mechanisms and controlling factors view project soils in the global agenda-ITPS-GSP? FAO view project, Chap. 3 (2017). [Online]. Available: https://www.researchgate.net/publication/319206365
18. V. Vijay et al., Review of large-scale biochar field-trials for soil amendment and the observed influences on crop yield variations. Front. Energy Res. **9**, 710766 (2021). https://doi.org/10.3389/FENRG.2021.710766/XML/NLM
19. R.P. Premalatha et al., A review on biochar's effect on soil properties and crop growth. Front. Energy Res. **11**, 1092637 (2023). https://doi.org/10.3389/FENRG.2023.1092637/XML/NLM
20. A. El-Naggar et al., Biochar application to low fertility soils: a review of current status, and future prospects. Geoderma **337**, 536–554 (2019). https://doi.org/10.1016/J.GEODERMA.2018.09.034
21. T.J. Purakayastha et al., A review on biochar modulated soil condition improvements and nutrient dynamics concerning crop yields: pathways to climate change mitigation and global food security. Chemosphere **227**, 345–365 (2019). https://doi.org/10.1016/J.CHEMOSPHERE.2019.03.170
22. S. Sharma et al., Biochar as a potential nutrient carrier for agricultural applications. Curr. Pollut. Rep. **11**(1), 1–36 (2025). https://doi.org/10.1007/S40726-025-00349-7
23. W. Christel, S. Bruun, J. Magid, W. Kwapinski, L.S. Jensen, Pig slurry acidification, separation technology and thermal conversion affect phosphorus availability in soil amended with the derived solid fractions, chars or ashes. Plant Soil **401**(1–2), 93–107 (2016). https://doi.org/10.1007/S11104-015-2519-0/FIGURES/2

24. L. Leng et al., Nitrogen containing functional groups of biochar: an overview. Bioresour. Technol. **298**, 122286 (2020). https://doi.org/10.1016/J.BIORTECH.2019.122286
25. M.I. Al-Wabel, A. Al-Omran, A.H. El-Naggar, M. Nadeem, A.R.A. Usman, Pyrolysis temperature induced changes in characteristics and chemical composition of biochar produced from conocarpus wastes. Bioresour. Technol. **131**, 374–379 (2013). https://doi.org/10.1016/J.BIORTECH.2012.12.165
26. J.M. De la Rosa, M. Paneque, I. Hilber, F. Blum, H.E. Knicker, T.D. Bucheli, Assessment of polycyclic aromatic hydrocarbons in biochar and biochar-amended agricultural soil from Southern Spain. J. Soils Sediments **16**(2), 557–565 (2016). https://doi.org/10.1007/S11368-015-1250-Z/FIGURES/3
27. Y. Ding et al., Biochar to improve soil fertility. A review. Agron. Sustain. Dev. **36**(2), 1–18 (2016). https://doi.org/10.1007/S13593-016-0372-Z
28. K. Mohammadi, J. Purkaystha, S.O. Prasher, M.T. Afzal, J. Dhiman, C. Nzediegwu, Effect of biochar, hydrogel and soil liming on nutrient leaching in a coarse-textured soil. Sustainability **17**(12), 5396 (2025). https://doi.org/10.3390/SU17125396
29. B. Glaser, J. Lehmann, W. Zech, Ameliorating physical and chemical properties of highly weathered soils in the tropics with charcoal—a review. Biol. Fertil. Soils **35**(4), 219–230 (2002). https://doi.org/10.1007/S00374-002-0466-4/METRICS
30. M. Wang, X. Yu, X. Weng, X. Zeng, M. Li, X. Sui, Meta-analysis of the effects of biochar application on the diversity of soil bacteria and fungi. Microorganisms **11**(3), 641 (2023). https://doi.org/10.3390/MICROORGANISMS11030641/S1
31. X. Zhang, X. Feng, N. Chai, Y. Kuzyakov, F. Zhang, F.M. Li, Biochar effects on crop yield variability. Field Crops Res. **316**, 109518 (2024). https://doi.org/10.1016/J.FCR.2024.109518
32. S.H. Bai et al., Combined effects of biochar and fertilizer applications on yield: a review and meta-analysis. Sci. Total Environ. **808** (2022). https://doi.org/10.1016/j.scitotenv.2021.152073
33. L. Zhong et al., Biochar reduces N_2O emission from fertilized cropland soils: a meta-analysis. Carbon Res. **4**(1), 1–14 (2025). https://doi.org/10.1007/S44246-025-00198-5

Chapter 4
Application of Biomass Ash and Biochars as Fertilizers

4.1 Introduction

Biomass ash and biochar are gaining increasing interest as alternative or supplementary fertilizers in agriculture and forestry. Both products are made from plant materials. Biomass ash through full combustion of biomass, while biochar through incomplete combustion (pyrolysis) under oxygen-limited conditions. The use of these products fits in with the idea of a closed-loop economy. It allows nutrients to be recycled from plant waste back into the soil, reducing the use of conventional mineral fertilizers and landfill. The following chapter discusses methods for the application of biomass ash and biochar to soil, reviews research results on their effectiveness and provides examples of practical use in different types of crops.

4.2 Application Methods

Biomass ash is most commonly used in loose or granulated form as a mineral fertilizer and liming agent. Due to its high alkaline content (oxides and carbonates of calcium, potassium, magnesium), it has a strongly alkaline reaction (pH often > 10) and can rapidly raise the pH of acidic soils. For this reason, ash is sometimes used as a substitute for lime to deacidify soils and forest peatlands [1, 2]. Prior to application, ash is often granulated or pelletized to facilitate spreading and reduce dust. In Scandinavia, for example, technologies have been developed to produce wood ash pellets specifically for forest fertilization in poor peat habitats [3]. In agriculture, ash can be spread with lime or fertilizer spreaders and then lightly mixed into the soil. Typical application rates depend on the properties of the ash and the soil [2]. It is important to avoid excessive doses, as very high single applications of ash can excessively raise the salinity and pH of the soil, leading to temporary stunting of plant growth and loss of potassium by leaching [4]. Biomass ash is rich in K, Ca,

K. Chojnacka et al., *Sustainable Fertilizers: Utilizing Biomass Ashes and Biochars*, SpringerBriefs in Applied Sciences and Technology,
https://doi.org/10.1007/978-3-032-03765-7_4

Mg, P and Si but contains virtually no nitrogen; its liming efficiency is high, with a neutralizing value of ≈ 45–55% $CaCO_3$-equivalent for typical wood ash [1]. The composition of the ash depends on the type of biomass burned, e.g. cereal straw ash contains more silicon and wood ash more calcium. Properly selected ash, free of impurities, can successfully serve as a potassium and calcium fertilizer to close the nutrient cycle in the agricultural system. Returning the ash back to the field allows a significant proportion of the elements previously taken up by plants to be returned to the soil, preventing soil depletion [5].

An important aspect is the quality of the ash. It must come from clean biomass (wood, straw, and vegetable waste) not showing excessive concentrations of heavy metals. Biomass ash may contain trace amounts of metals (Cd, Zn, Pb, Cu, etc.), especially if the fuel was wood from long-growing trees accumulating these elements [1]. Therefore, the ash should be tested for contaminants before being used in agriculture. In the European Union, the use of ash is regulated, inter alia, through the regulation of fertilizer products, namely Regulation (EU) 2019/1009, which opened the way for the certification of certain ashes as fertilizers, provided that standards (including heavy metal limits) are met [6]. Ash that fails to meet these limits remains a 'waste' under Directive 2008/98/EC and must be managed accordingly. Ash can also be enriched or mixed with other materials prior to application. For example, mixtures of ash with organic waste (manure, compost) are being investigated to supplement nitrogen deficiency and increase fertilizer efficiency [5]. An interesting innovation is ash-biochar composites, i.e. materials obtained by adding ash to biomass before pyrolysis, resulting in an ash-saturated biochar. Such a product combines the porosity of biochar with the mineral richness of ash. Studies have shown that ash-blended biochar releases nutrients more slowly than pure ash (lower conductivity and less potassium leaching), while immobilizing some of the toxic heavy metals (e.g., chromium) from the ash. The use of such composites reduces the risk of salinity and toxicity, improving the safety of ash fertilization [4].

Biochar is a porous black carbon produced from biomass at 300–700 °C. Unlike ash, biochar contains mainly stable organic carbon and the content of directly available minerals is sometimes lower (many elements remain in mineral form in ash after complete combustion, whereas in biochar some may be bound in organic structures or volatilized during pyrolysis). Nevertheless, biochar often contains some phosphorus, potassium, calcium, and micronutrients. Especially if the starting biomass was rich in these elements. Unlike ash, biochar has a neutral to slightly alkaline reaction (typically pH ~ 7–9, depending on the raw material and pyrolysis temperature). Due to its highly developed specific surface area and porosity, the biochar acts primarily as a soil improver. It improves water retention, increases the sorption and cation capacity of the soil, promotes the development of beneficial soil microflora and reduces fertilizer losses through leaching [1, 7]. It is very important to enrich biochar with nutrients before application, as fresh, pure biochar may initially sorb available fertilizers (especially nitrogen) in the soil, making them less available to plants [8]. In practice, it is often recommended to mix biochar with manure, slurry, compost or mineral fertilizers and wait some time before adding it to the soil. Such enriched biochar then acts as a carrier that slowly releases nutrients while improving soil structure.

4.3 Experiments and Field Trials

The effectiveness of biomass ash and biochar as fertilizers has been the subject of numerous studies, both in pot experiments and field trials on different soils and crops. The results of these studies indicate that moderate doses of ash or biochar can have a beneficial effect on soil fertility and yield, although the effects depend on many factors, such as the dose and composition of the material, soil type, water conditions, crop type, and concurrent nitrogen fertilization.

Greenhouse trials show that modest additions of biomass ash can raise soil pH and supply exchangeable cations, changes that often enhance plant performance. An unpublished tobacco study attributed to Wang et al. reported that incorporating 0.2% w/w sieved wood ash into an acidic loam lifted the pH from 5.94 to 6.35 and increased extractable K by about 30%. Treated plants formed larger leaves, sustained higher net photosynthetic rates and developed a denser network of lateral roots than the unfertilized control. Indole-3-acetic acid concentrations in both leaves and roots rose, pointing to hormonal stimulation of root growth, and the accompanying gain in total biomass indicated more efficient CO_2 assimilation [9].

Other experiments have shown that even though ash lacks nitrogen, its application can increase soil nitrogen availability. Johansen et al. conducted a 4.5-month experiment on an acid poor forest soil with grass (*Deschampsia flexuosa*), adding different ash doses (0, 1.1, 3.3, 11 and 33 t ha^{-1}). The ash clearly stimulated plant biomass growth, even more than the corresponding doses of pure CaO (lime) or potassium-phosphorus fertilizers, suggesting a synergistic effect of pH raising and nutrient supply. Despite the very low N content of ash, increased nitrogen uptake by plants was observed in plots with ash relative to the control. This was explained by the fact that the ash, through deacidification and the addition of readily available bases and phosphorus, stimulated the mineralization of soil organic matter, releasing additional nitrogen for the plants. This means that biomass ash can indirectly improve the nitrogen nutrition of crops, as long as the soil contains organic matter capable of mineralization. In the cited experiment, only the highest, extremely high ash dose (33 t/ha) caused some negative effects, including increased cadmium accumulation in plants. Doses up to 11 t/ha did not cause a significant increase in the heavy metal content of the plant parts, even though the ash used contained quite high levels of Cd (16.3 mg/kg). This is an important finding indicating that at moderate doses the risk of contamination of the food chain with metals from ash is low [2].

Many field studies have looked at replacing or supplementing traditional mineral fertilizers with biomass ash. For example, Cruz-Paredes et al. investigated the use of wood ash as a replacement for phosphate mineral fertilizer in arable fields. They found that ash could partially replace superphosphate, keeping yields similar while not causing excessive cadmium accumulation in plants [10]. A positive effect of ash on the development of mycorrhizal fungi in the soil was observed, suggesting its beneficial effect on the biological properties of the soil (due to the increase in pH and the addition of Ca, P). Forestry experiments with ash fertilization on poor soils and peatlands have been carried out in Finland and Sweden for decades. These have shown that wood ash is effective in increasing the content of basic cations in the soil, reducing soil acidification and increasing the weight gain of trees on clearcuts. In Canada, ash application in forests can mimic the positive effects of fires (which return ash to the soil) by counteracting the depletion of intensively managed forest habitats [3].

Biochar has also been intensively studied for its effects on yields and soil properties. Meta-analyses of experimental results indicate that the addition of biochar promotes yield increases, especially on soils with low initial fertility [7, 11]. For example, a meta-analysis of 26 pooled studies unequivocally showed a beneficial effect of biochar on various agronomic parameters, from yield to root biomass to soil water holding capacity [7]. On average, biochar was found to increase yields by about 10–15% overall, with the effects being more pronounced when biochar was applied together with fertilizer. Ye et al. separately analyzed the results of experiments comparing: (a) control without fertilizer vs. biochar, and (b) control with full fertilization vs. biochar + fertilization. In the first case, the average yield increases due to biochar reached ~ 10%, while in the second case, where both the control and the biochar objects received fertilizers—the addition of biochar gave a further yield increase of ~ 15% relative to fertilization alone [8]. The implication is that biochar does not replace traditional nitrogen or phosphorus fertilization, but acts as a factor to improve the efficiency of fertilizer use by plants. Importantly, meta-analyses also indicate that biochar works most effectively on degraded, acidic, sandy, or clay soils with low organic matter content, where water and nutrient deficits are greatest. Under these conditions, biochar additions produce the greatest yield gains. In contrast, in fertile, rich and high pH soils, the effects are sometimes smaller or negligible (e.g., no effect of biochar on rice yield was found in flooded soils with neutral pH) [11].

An example of a specific effect of biochar is its effect on greenhouse gas emissions from soil. Several studies have confirmed that biochar additives can reduce nitrous oxide (N_2O) emissions from agricultural soils by adsorbing ammonium nitrogen and creating more favorable conditions for N immobilization in the soil [12]. In contrast, the effect on CO_2 emissions from soil is dose-dependent. Biochar contains stable carbon, which itself does not rapidly mineralize, but can stimulate or inhibit the decomposition of existing soil matter. The carbon balance of the soil improves after biochar application, which is beneficial in terms of CO_2 uptake from the atmosphere [7].

During research on biomass ash and biochar, attention has also been paid to environmental safety. Field trials monitor, among other things, the content of heavy metals in plants and soils, as well as possible effects on groundwater. The results so far are generally reassuring: when using materials that meet the standards (with low metal content), no exceedances of the permissible concentrations of heavy metals in agricultural crops or significant groundwater contamination are observed [2]. For example, water quality studies in the vicinity of surface-fertilized forests (wood ash vs. nitrogenous fertilizers) in Latvia showed no negative impact of ash on the physicochemical and biotic parameters of waters draining from the forest catchment. There was no increase in salinity or nutrients in streams after ash application [13]. This does not mean that risks do not exist. Caution is needed, especially in the vicinity of surface waters (avoiding over-fertilization with ash which can increase alkaline run-off). For biochar, the risk of introducing polycyclic aromatic hydrocarbons (PAHs), which can be formed during pyrolysis, is being investigated [14].

4.4 Examples of Applications in Different Types of Crops

4.4.1 Field Crops

In field farming, biomass ash has been tested on a variety of crops, including cereals, maize, oilseed rape, fodder crops and biomass energy crops. The ash is treated here as a potassium-calcium and phosphorus fertilizer, which can partially replace conventional compound fertilizers. An example of a successful application is maize (*Zea mays*). Romdhane et al. conducted a field and pot experiment in which they introduced 0.1% (w/w) wood ash into a silty soil before sowing two maize varieties (drought-tolerant and drought-sensitive). This ash contained significant amounts of Ca, K, Mg, P, and S, although elevated levels of Zn, Pb, Cu (presumably from wood contamination) were also recorded. The results showed that the addition of ash increased the growth of maize stalks and the number of maize leaves, while grain yield remained at a similar level to the control. The most interesting effect, was the improvement in drought tolerance of maize—plants in plots with ash were slower to reduce transpiration during periods of water shortage, maintaining photosynthetic activity for longer. In particular, the tolerant variety (D24) with ash addition developed a deeper root system (reaching 1.5 m), which allowed it to make better use of ground water. The authors conclude that wood ash can be successfully used in maize fertilization to improve growth and drought tolerance, but attention should be paid to the quality of the ash (avoid ash contaminated with metals to avoid introducing them into the food chain) [15].

Cereals are another example. Cereal and cereal-corn straw ash has been tested as a local fertilizer on wheat and barley fields. Although this ash contains less calcium than wood ash, it still provides a lot of potassium and phosphorus. In Poland, Piekarczyk et al. analyzed the composition of ashes from straw and hay burning. They found that they had a favorable ratio of basic macronutrients and could provide a full-value potassium and calcium fertilizer for light soils. The main drawback is elevated chloride; when Cl^- exceeds $\approx 0.5\%$ DM, rates must be reduced or mixed with organic matter to protect Cl-sensitive crops such as potato. In field experiments with barley fertilized with straw ash, there was a neutral or slightly positive effect on grain yield, but a clear improvement in soil pH and abundance of K, Mg and P after several years of application. This means that ash can gradually enrich the soil, although the yield-forming effect may only become apparent after some time [16].

Energy crops such as perennial grasses (miscanthus, sida hermaphrodita) or fast-growing trees (energy willow) can also benefit from fertilization with ash and biochar. These crops are grown to produce biomass for energy purposes, so it is natural to return the ash from their combustion to the field with these plants. This closes their biogeochemical cycle. An example is sugar sorghum, an annual crop with high biomass potential. Romanowska-Duda et al. studied a sorghum crop in which sorghum ash (obtained from burning the previous crop) was added to the soil in combination with biogas plant digestate and watercress extract. The aim of such a combination was an integrated yield increase and CO_2 sequestration. The results showed that sorghum responded very well to ash fertilization. There was an increase in biomass yield and accumulation of additional organic carbon in the soil. Biostimulant additives (digestate, extract) further increase the efficiency of the ash fertilizer. This approach points in the direction of sustainable intensification: combining ash (which provides minerals) with biochar or other organic preparations to simultaneously improve soil fertility and reduce greenhouse gas emissions [5].

A number of positive results have been reported for biochar in field crops, especially when growing crops on poor soils. One of the best documented is the cultivation of rice on tropical soils, where the addition of bamboo biochar improved the structure of cracked clay soil and increased yields by several per cent, while reducing methane emissions from flooded fields (the biochar acted as an oxidant and adsorbent) [12, 17]. In maize crops, meta-analyses indicate an average yield increase of about 20% after biochar. In wheat and other cereal crops, the effects are somewhat smaller (in the order of 10–15%), which is related to the fact that shallow-rooting cereals do not always fully benefit from the improvement of deeper soil layers by biochar. Very promising are the results for oilseed crops (e.g. rape, sunflower) and root crops, here biochar can give significant increases in biomass and seed yield. A meta-analysis by Xu et al. showed that the greatest relative benefits of biochar were observed for oilseed crops (+ 37% yield) and vegetables (+ 28%), compared to + 12–26% for cereals. This is explained by the fact that vegetables, for example, are often grown intensively on soils that require improved structure and water holding capacity, and biochar provides the greatest benefits in these aspects [11].

4.4.2 *Forestry Crops*

The use of biomass ash in forests aims to replenish nutrients removed from the ecosystem during timber harvesting and to neutralize the acidification of forest soils. In Scandinavian countries, fertilization of forests with mineral fertilizers containing nitrogen was practiced from the 1950s to increase tree growth, but routine nitrogen fertilization was abandoned in the 1990s for ecological reasons [13]. There is now a greater emphasis on fertilization with wood ash, especially on clearcuts on peaty and nutrient-poor soils (where intensive biomass harvesting leads to habitat depletion). Wood ash functions chiefly as a liming-plus-potassium amendment: it delivers large doses of Ca (often 20–40% CaO) and several percent K_2O, while its plant-available phosphorus is modest, typically below 2% P_2O_5, so it should be regarded only as a supplementary P source. In Finland, for example, tens of thousands of tons of ash from biomass CHP plants are used annually to fertilize forests, mainly on peatland clearcuts. The granulated ash results in a significant increase in the content of bioavailable P, K, and Ca in the soil and an increase in pH, which creates more favorable conditions for the development of the root system of trees. A Swedish study found that ash fertilization on peatland logging plots significantly increases the biomass gain of trees (spruce, pine) compared to unfertilized controls. This is especially the case where the soils were strongly acidic and poor before treatment—the ash there acts in a similar way to past natural fires, which provided ash and increased the fertility of the habitat. An additional benefit is the stabilization of pH. This counteracts soil acidification caused by decomposing needles and acid rain [3].

Biochar is also used in forests, although on a smaller scale than ash. Wood biochar is sometimes added to the soil when afforesting degraded areas or as a fertilizer ballast component in forest nurseries. Its role is mainly to improve water retention and air–water conditions for tree seedlings, as well as long-term carbon fixation in the forest ecosystem. For example, in a study by Santalla et al. applied wood ash containing charcoal (produced by incomplete combustion) to a pine plantation, resulting in improved tree growth due to a combined fertilizing and conditioning effect (the charcoal in the ash improved the soil structure). It can therefore be said that in forestry, the combination of ash and biochar gives the best results: the ash provides nutrients, while the biochar ensures better retention of these nutrients in the root zone and improves soil moisture. There are currently pilot projects to add biochar to the soil before afforestation of wasteland and to use biochar as an additive to nursery substrate (for the production of more resistant seedlings) [5, 15].

Different types of crops benefit from the use of biomass ash and biochar in different ways. In forests, it is mainly to replenish calcium, magnesium and phosphorus and to counteract acidification after logging (ash) and to improve water and soil conditions (biochar). In field crops, increasing the abundance of K, P, Ca (ash) and improving the structure and water capacity of the soil (biochar), resulting in higher and more stable yields of cereals, maize or energy crops. The specifics of the crop and soil must always be taken into account. Used appropriately, biomass ash and biochar provide a valuable, low-carbon and sustainable source of fertilizer, contributing to both closing the nutrient cycle and increasing the resilience of agriculture to climate change by improving soil quality.

References

1. L. Romdhane et al., Effects of soil amendment with wood ash on transpiration, growth, and metal uptake in two contrasting maize (*Zea mays* l.) hybrids to drought tolerance. Front. Plant Sci. **12**, 661909 (2021). https://doi.org/10.3389/FPLS.2021.661909/BIBTEX
2. J.L. Johansen et al., The complexity of wood ash fertilization disentangled: effects on soil pH, nutrient status, plant growth and cadmium accumulation. Environ. Exp. Bot. **185** (2021). https://doi.org/10.1016/J.ENVEXPBOT.2021.104424
3. K.D. Hannam et al., Wood ash as a soil amendment in Canadian forests: what are the barriers to utilization? Can. J. For. Res. **48**(4), 442–450 (2018). https://doi.org/10.1139/CJFR-2017-0351
4. W. Buss, S. Jansson, O. Mašek, Unexplored potential of novel biochar-ash composites for use as organo-mineral fertilizers. J. Clean. Prod. **208**, 960–967 (2019). https://doi.org/10.1016/J.JCLEPRO.2018.10.189
5. Z. Romanowska-Duda, R. Janas, M. Grzesik, Increasing fertilization efficiency of biomass ash by the synergistically acting digestate and extract from water plants sequestering CO_2 in sorghum crops. Molecules **29**(18), 4397 (2024). https://doi.org/10.3390/MOLECULES29184397
6. EP, Regulation (EU) 2019/1009 of the European Parliament and of the Council of 5 June 2019 Laying Down Rules on the Making Available on the Market of EU Fertilising Products and Amending Regulations (EC) No 1069/2009 and (EC) No 1107/2009 and Repealing Regulation (EC) No 2003/2003 (2019)
7. H.P. Schmidt et al., *Biochar in Agriculture—A Systematic Review of 26 Global Meta-Analyses* (Wiley, 2021). https://doi.org/10.1111/gcbb.12889
8. L. Ye, M. Camps-Arbestain, Q. Shen, J. Lehmann, B. Singh, M. Sabir, *Biochar Effects on Crop Yields with and Without Fertilizer: A Meta-Analysis of Field Studies Using Separate Controls* (Blackwell Publishing Ltd., 2020). https://doi.org/10.1111/sum.12546
9. R. Wang et al., Biomass ash as soil fertilizers: supercharging biomass accumulation by shifting auxin distribution. Chemosphere **357** (2024). https://doi.org/10.1016/j.chemosphere.2024.141910
10. C. Cruz-Paredes, Á. López-García, G.H. Rubæk, M.F. Hovmand, P. Sørensen, R. Kjøller, Risk assessment of replacing conventional P fertilizers with biomass ash: residual effects on plant yield, nutrition, cadmium accumulation and mycorrhizal status. Sci. Total Environ. **575**, 1168–1176 (2017). https://doi.org/10.1016/j.scitotenv.2016.09.194
11. Z. Xu, R. Zhou, G. Xu, Global analysis on potential effects of biochar on crop yields and soil quality. Soil Ecol. Lett. **7**(1), 240267 (2025). https://doi.org/10.1007/S42832-024-0267-X
12. L. Zhong et al., Biochar reduces N_2O emission from fertilized cropland soils: a meta-analysis. Carbon Res. **4**(1), 1–14 (2025). https://doi.org/10.1007/S44246-025-00198-5
13. D. Ozolinš et al., The impact of forest fertilization on the ecological quality of two hemiboreal streams. Forests **13**(2), 196 (2022). https://doi.org/10.3390/F13020196/S1
14. J.M. De la Rosa, M. Paneque, I. Hilber, F. Blum, H.E. Knicker, T.D. Bucheli, Assessment of polycyclic aromatic hydrocarbons in biochar and biochar-amended agricultural soil from Southern Spain. J. Soils Sediments **16**(2), 557–565 (2016). https://doi.org/10.1007/S11368-015-1250-Z/FIGURES/3
15. M. Santalla, B. Omil, R. Rodríguez-Soalleiro, A. Merino, Effectiveness of wood ash containing charcoal as a fertilizer for a forest plantation in a temperate region. Plant Soil **346**(1), 63–78 (2011). https://doi.org/10.1007/S11104-011-0794-Y
16. M. Piekarczyk, K. Kotwica, D. Jaskulski, The elemental composition of ash from straw and hay in the context of their agricultural utilization. Acta Sci. Pol. Agric. **10**(2), 97–104 (2011)
17. J.M. Lee et al., Effects of biochar on methane emissions and crop yields in East Asian paddy fields: a regional scale meta-analysis. Sustainability **15**(12), 9200 (2023). https://doi.org/10.3390/SU15129200

Chapter 5
Impact on Crop Yield and Quality

5.1 Introduction

Recent energy-balance studies put worldwide biomass ash production at roughly 40–60 million tons per year, not hundreds of millions [1, 2]. Because wood- or straw-derived ash is rich in K, Ca, Mg, P, and micronutrients, it can be reused as a nutrient input in circular agriculture; yet trace Cd, Pb and Hg may still occur, so every batch should be tested to confirm it meets fertilizer-quality limits [3]. Biochar is a highly carbonized material obtained in controlled pyrolysis of biomass (with limited oxygen). It is characterized by high porosity, high specific surface area, and an alkaline reaction [4]. Biochar is valued mainly for its ability to improve soil fertility and sequester carbon [5] (Fig. 5.1).

5.2 Yield Effects

5.2.1 Cereal Crops

Numerous studies have documented the positive effects of ash on cereals, especially on acidic and potassium- or phosphorus-poor soils. Biomass ash is sometimes considered as a substitute for phosphorus and potassium fertilizers in sustainable agriculture. For example, in Danish experiments, replacing traditional superphosphate with biomass ash in a spring barley crop kept grain yields unchanged [6]. There were no negative effects in the form of heavy metal accumulation—the cadmium content in grain remained unchanged, and the total productivity of the crop was comparable to the control fertilized with conventional fertilizer [6]. Biomass ash provides valuable nutrients for cereals (including P, K, Ca, Mg), deacidifies the soil and can stimulate microbial activity, resulting in better nutrition and plant growth. For example, spring

K. Chojnacka et al., *Sustainable Fertilizers: Utilizing Biomass Ashes and Biochars*, SpringerBriefs in Applied Sciences and Technology,
https://doi.org/10.1007/978-3-032-03765-7_5

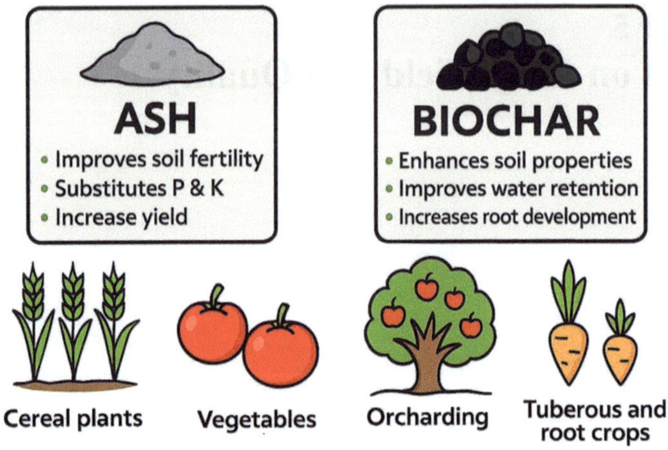

Fig. 5.1 Fertilizing effects of biomass ash and biochar on key plant groups

barley grown on light soil with the addition of ash (at rates that provide an equivalent amount of P and K as traditional fertilizers) achieved more than double the biomass increase compared to the control [7]. The application of ash also had a beneficial effect on the quality of barley grain—higher content and better protein composition were recorded after ash fertilization. In field experiments with wheat, phosphorus applied in the form of wood ash proved to be as effective as traditional calcium phosphate in increasing grain yields [8]. Positive effects were also noted for other annual cereal crops, including in corn and buckwheat, the introduction of biomass ash (as a source of P and K) yield increases comparable to or higher than traditional mineral fertilization [9]. Moderate ash doses (< 6 t ha^{-1}) supply K, P, and Ca and raise soil pH, but-because the ash is virtually N-free-they must be paired with the standard mineral-N topdressing for cereals ($\approx$ 80–120 kg N ha^{-1}, depending on variety and yield goal) to avoid N deficiency.

In contrast to ash, biochar itself is not a rich mineral fertilizer—it mainly contains stable carbon and small amounts of ash (depending on the raw material). Rather, its effect on cereals manifests itself indirectly, by improving soil properties: increasing water holding capacity, cation exchange capacity (CEC), creating a favorable habitat for microorganisms and earthworms, and raising pH on acidic soils [5]. Meta-analyses of field studies indicate that biochar application alone (without fertilizers) usually does not significantly increase grain yields, especially on fertile soils [10]. Biochar combined with traditional fertilization has a synergistic effect—it improves the efficiency of nutrient utilization by plants, leading to an additional increase in yield over the level obtained from mineral fertilizers alone [10]. For example, a meta-analysis of 50 studies showed that the addition of biochar to standard NPK increases grain yields by an average of about 15% compared to the sole application of NPK In experiments on wheat grown on poor sandy soil, a single application of biochar resulted in a gradual increase in grain yield in subsequent years—after 3 years, yields were about 30% higher than in control plots [10]. This effect is explained by the improvement of soil sorptive properties and pH by biochar (the so-called liming effect), as well as the long-term increase in soil organic carbon content. The response of cereals to biochar depends on soil and climatic conditions. On fertile, nutrient-rich soils, the effects may be minimal or delayed. In contrast, on acidic and poor soils (such as in the tropics), the greatest yield increases are recorded after biochar application [11]. For instance, analysis of several hundred experiments showed that in tropical climates, biochar significantly increases yields, while in temperate climates the effects were statistically weaker or inconclusive [4]. Biochar is a promising addition to the cereal fertilization system, especially where improved water retention and soil deacidification are desired, but the most favorable results are achieved by combining it with conventional fertilizers.

5.2.2 *Tuberous and Root Crops*

Root crops (such as potato, sugar beet, cassava, yam, and carrots) have potassium and calcium requirements, making biomass ash a potentially attractive fertilizer for these crops. In Poland, biomass ash has been tested in food and starch potato crops, among others, with promising results. When applied at a moderate dose, the ash not only increased tuber yield, but also improved tuber quality and technological characteristics [12, 13]. For example, a significant increase in the resistance of potato tubers to mechanical damage after ash fertilization was found, which is an important quality parameter that affects harvesting and transport losses [13]. In fact, ash, by raising the soil reaction and providing Ca, reduces the tubers' susceptibility to scab-type diseases and improves skin structure. Other experiments have proven that ash can be successfully used as a source of potassium for root crops. In sugar beet cultivation, ash from vegetable pomace replacing KCl promoted sugar accumulation in roots and did not contaminate soils with heavy metals. Root crops are sensitive to salinity, and too high a single dose of ash can raise the salt concentration in the root zone and impair emergence. Studies on test plants indicate that exceeding a certain threshold of ash dose begins to have a depressive effect on the growth of root crops. For example, in an experiment with willow, the addition of ash at doses equivalent to > 3 t/ha resulted in reduced biomass growth, especially when the ash was not well mixed into the soil [14]. Thus, optimizing the application rate and possible gradual applications are key. Moderate fertilization with biomass ash can increase tuber and root yields by improving potassium, phosphorus, and calcium nutrition and raising soil pH, which promotes better binding of potato tubers and beet roots.

Biochar alone raises tuber yields by only $\approx 5\%$, but when co-applied with mineral fertilizer the mean gain rises to $+\ 18\%$ (global meta-analysis of 62 comparisons), mainly to improve the properties of light soils and increase fertilizer efficiency. Results show measurable benefits in both yield and crop quality. For example, in field experiments in Bangladesh, the addition of biochar to potato-grown sandy soil significantly increased tuber yield (depending on the dose by 15–30% relative to the control) and improved tuber quality parameters such as dry matter content and specific gravity of flesh [15]. An increase in the concentration of dry matter in the tubers suggests better technological quality (more mealy potatoes, beneficial for the starch industry, for example). In this study, biochar applied with mineral fertilizers proved optimal—e.g., half of the recommended dose of fertilizers and biochar at 5 t/ha gave yields equal to or exceeding those with the full dose of fertilizers without biochar [15]. Biochar retains more water and nutrients in the soil, which is beneficial for root crops with shallow root systems sensitive to drought [16]. Other experiments have shown that biochar can also improve the quality of root crops in terms of nutritional value [17]. Unlike cereals, root crops often respond well to improvements in soil structure and moisture, so the effects of biochar can sometimes be seen more quickly here. Too high doses of biochar may hinder the germination of small seeds or lead to a temporary depletion of nitrogen in the soil [18]. Therefore, it is recommended to combine biochar with nitrogen fertilizers or nitrogen-rich organic matter (e.g. compost, manure), which protects against possible immobilization effects.

5.2.3 Vegetables

Vegetables, especially species with short growing seasons and shallow root systems, are highly sensitive to soil nutrient abundance and soil pH. A global meta-analysis found that soil acidification reduces yields by an average of about 14%, with vegetable yields falling more sharply than those of cereals or legumes [19]. Biomass ash, thanks to its strong deacidifying properties and abundance of easily soluble calcium and potassium, is sometimes used in vegetable cultivation to improve the parameters of acidic soils. For example, vegetables such as tomatoes and peppers grown on acidic soils respond positively to skillful liming with wood ash, which raises the pH in the root zone and provides calcium—an element crucial to fruit quality. The ash also contributes potassium, which is essential for fruit ripening. Thus, wood ash can act as a fast-acting lime while enriching the soil with K, Ca, and Mg [20]. In greenhouse experiments, it was shown that fertilizing tomatoes with biomass ash allowed yields comparable to mineral fertilization, while maintaining fruit quality [21]. Under field conditions, a single application of a moderate dose of ash before crop establishment increased soil pH and potassium availability, resulting in higher yields and better vegetable quality. For example, it was observed that cabbage heads were heavier and more compact after the ash application compared to the control on acidic soil [22]. Reports of extreme one-off doses ($\geq$ 100 t ha^{-1}) are not representative; agronomic guidelines recommend 2–10 t ha^{-1} of qualifying ash to avoid excessive salinity, alkalinity and metal loading: cabbage, onions, tomatoes or lettuce [23]. Although such a dose is extremely high, it shows the potential of ash to rehabilitate very poor soils. More typical doses (5–10 t/ha) also bring tangible benefits: for example, the addition of 5 t/ha of ash to sandy soil for onion cultivation clearly increased the weight of the above-ground part (leaves), indicating better nutrition and vitality of the plants [24]. A Polish pot study reported that wood-ash doses of 5 t ha^{-1} produced chard biomass comparable to mineral fertilizer treatments (specific reference required). Root vegetables (e.g., carrots, parsley) can accumulate heavy metals—so when fertilizing with ash, it is important to control the dosage and composition of the ash (e.g., avoiding ash from burning mixed waste, which can contain a lot of Pb, Cd). Ash from the combustion of pure biomass generally has low levels of these pollutants and studies show that it does not cause exceedances in vegetable yields [6].

The use of biochar in vegetable horticulture is being intensively researched because vegetables, due to their shallow rooting and short development cycle, respond quickly to changes in soil properties. Biochar improves the structure and aeration of the soil, increases its water holding capacity and buffers the availability of nutrients, which promotes a continuous supply of water and fertilizers to plants during critical growth phases [25]. On sandy soils, the addition of biochar often results in better water supply for vegetables during periods of rainfall deficiency due to increased moisture retention. For example, in a field experiment in Ghana, the application of biochar from corn husks at a rate of 5 t/ha along with 50% of the standard NPK fertilizer rate significantly increased the size and weight of cabbage heads compared to mineral fertilizer alone [26]. There, more pronounced increases in plant height and leaf area were also observed with the combination of biochar and fertilizer, explaining the higher marketable yield. In addition, better uptake of nitrogen, phosphorus and potassium by cabbage plants was observed in the presence of biochar [26]. Similar effects were noted for leafy vegetables such as lettuce and spinach—biochar increased leaf biomass and chlorophyll content, indicating better nitrogen nutrition [25]. The results for fruit vegetables (tomato) under drought stress are particularly promising: the addition of biochar improves water retention, reduces moisture fluctuations in the root zone, resulting in less yield loss during periods of water shortage [27]. For example, tomatoes from crops with biochar had higher levels of sugars and vitamin C in the juice, which was linked to better potassium nutrition and less "water stress" on the plants [28].

5.2.4 Orcharding

Orcharding is an area where biochar is gaining interest, although it requires a specific strategy: a single soil application intended to deliver long-term benefits. Biochar may be especially helpful in older, depleted orchards where soil structure has deteriorated and organic-matter levels have fallen. A good example comes from apple orchards on China's Loess Plateau (sometimes called the Lesser Highlands), where more than 80% of orchards are now over 20 years old and suffer yield decline and poorer fruit quality because of soil fatigue. In one study at such a site, researchers incorporated wood-derived biochar at rates of 2–12 kg per tree (roughly 5–30 t ha^{-1}) and tracked changes in soil properties and apple yield for two years [29]. The best results were obtained at an average dose of ~ 6 kg of biochar per tree, the fruits from these trees were of the highest quality (size, coloration, and sugar content), and the yield per tree was the highest. Too high doses (10–12 kg per tree) did not further improve the results, and in some aspects even worsened them (e.g., too high pH reduced the availability of zinc and symptoms of zinc deficiency appeared on the leaves) [29]. This example shows that moderate use of biochar can effectively revitalize an old orchard, increasing its productivity and production quality by improving soil conditions. Other work also confirms the beneficial effects of biochar on fruit trees. A certain challenge in orcharding is the one-time introduction of a large amount of material (digging between rows of trees), but alternative methods of application are being practiced, such as deep pockets of biochar (drilling holes around the root zone and filling with a mixture of biochar and compost)—such techniques are being tested, for example, in apple orchards in the US, achieving improvements in tree growth and yield the very next season. For example, in Indonesia, mango plantations were established on soils enriched with biochar and an acceleration of the trees' entry into fruiting and an increase in yield was noted—fruit weight per tree was 26% higher at a biochar dose of 40 t/ha compared to the control [30]. Biochar is a promising additive in orchards and perennial plantations to improve soil conditions and thus benefit fruit yield and quality. It requires precise dosage selection and monitoring of pH and micronutrients to ensure optimal effect without side effects.

5.3 Comparison with Conventional Mineral and Organic Fertilizers

The ash from biomass combustion is characterized by an almost complete absence of nitrogen, while it contains significant amounts of phosphorus, potassium, calcium, and other nutritional elements needed by plants [31]. Due to the high concentration of bases (metal oxides and hydroxides), the ash has a strong deacidifying effect, acting as a natural liming agent [31]. Typical wood ash contains negligible nitrogen, approximately 2–5% P_2O_5, 3–7% K_2O, and 20–40% CaO. Its fertilizer value lies primarily in its phosphorus, potassium, and calcium content, rather than a balanced N:P:K ratio, offering virtually no nitrogen but serving as a useful source of phosphorus and potassium [32]. In comparison, conventional mineral fertilizers (e.g. NPK fertilizers) contain high concentrations of specific macronutrients (nitrogen, phosphorus, potassium) in inorganic forms (mineral salts) [33], while they do not provide organic matter or micronutrients (except in special mixtures). Organic fertilizers, such as compost or manure, have a lower NPK concentration, but contribute a diverse set of nutrients (including micronutrients) and a significant amount of organic carbon in the form of humus to the soil (Table 5.1).

The important differences are in the form and rate of release of nutrients from these fertilizers. The components of mineral fertilizers are water-soluble and immediately available to plants, which provides a quick effect, but also promotes loss through leaching (especially of nitrates) and requires frequent replenishment. In organic fertilizers, most of the nitrogen, phosphorus or sulfur is in organic forms that must

Table 5.1 Comparative characteristics of mineral, organic, ash, and biochar fertilizers

Feature	Mineral fertilizers	Organic fertilizers	Biomass ash	Biochar
NPK content	High, in readily available forms	Moderate, mostly in organic forms	Almost no N, high in P and K	Low, poorly available P and K
Organic matter	None	High (rich in humus)	None	High (mostly stable carbon)
Effect on soil pH	Can acidify soil (especially N fertilizers)	Slightly alkaline or pH-buffering	Strongly alkaline, raises soil pH	Alkaline, but weaker effect than ash
Nutrient release rate	Fast (immediate availability)	Slow (requires mineralization)	Fast for K and Ca, slower for P	Very slow; acts more as nutrient carrier
Effect on soil structure	None	Improves structure and water retention	Limited effect (mainly Ca addition)	Improves porosity and water holding
Effect on soil microbiota	Neutral or negative long-term	Strongly positive (carbon source)	Moderate (stimulates microbes via pH and nutrients)	Indirectly positive (habitat for microbes)
Yield-boosting efficiency	High, short-term effect	Moderate to high, long-term	High for P, K, Ca when combined with N	Limited alone, enhanced when combined with fertilizers
Soil persistence	Short (days to weeks)	Medium (months)	Medium (1–2 years for pH effect)	Long (years to decades)
Best use case	Quick nutrient supply, peak yields	Sustained fertility and soil health	Liming, supplementing K and P	Soil amendment, retention of nutrients

be mineralized—so their release is slower and spread out over time. For example, combining slow-release organic matter with a fast-release fertilizer (manure) made it possible, in a field experiment, to provide stable plant nutrition throughout the season [34]. Biomass ash contains minerals in the forms of oxides, carbonates and salts, so that some of them (such as potassium) are easily soluble and rapidly increase the availability of this element in the soil [35]. Studies confirm that wood ash can successfully act as a potassium fertilizer—it contains a lot of potassium and at the same time raises the pH of the soil [32]. Among other things, the phosphorus present in ash is in the form of calcium phosphates, making it less soluble than in superphosphate, but under soil conditions it gradually becomes available to plants (especially at elevated pH). Biochar, unlike ash, contains mainly stable elemental carbon and only a small ash mineral fraction—so by itself it provides plants with small amounts of bioavailable macronutrients. Some potassium or phosphorus may remain in biochar, but they are poorly soluble. In practice, therefore, biochar is not a typical fertilizer, but rather a carrier and sorbent for ingredients—it binds and retains externally supplied fertilizers, reducing their loss, and can release them gradually [36]. The fertilizing effect of biochar is mainly revealed when combined with conventional mineral or organic fertilizers, since biochar alone without the addition of readily available nitrogen usually does not increase yields, and in extreme cases fresh biochar with a high carbon content can even temporarily reduce nitrogen availability by immobilizing this element in the soil [37].

Mineral fertilizers have a quick but short-lived effect—the released ions are taken up in a few days or leached out, and no lasting trace of them remains in the soil. Organic fertilizers have a slower but longer-lasting effect: the organic matter built into them decomposes gradually, releasing components over many months. The nutritional effect of manure or compost is more evenly distributed over time and sustains soil fertility over the long term [34]. Ash releases a significant part of its components (especially potassium, calcium) immediately after application, giving an effect comparable to a single mineral fertilization with potassium and phosphorus. At the same time, its effect on the pH (raising the pH) can last for several seasons—similar to liming, depending on the application rate and soil properties. Biochar, is very persistent in the soil environment—the organic carbon it contains is in a form that is difficult to degrade. A single application of biochar can affect the soil for many years, especially in terms of improving the content of stable organic matter and associated physicochemical properties [36]. While biochar remains in the soil for a long time, its fertilizing effect requires the parallel application of other sources of ingredients—the effect of biochar itself is more long-term than short-term. Optimal fertilization strategies often involve a combination of these materials: rapid replenishment of NPK with mineral fertilizers, a slower source of nutrients and humus from organic matter, and long-lasting biochar to improve soil properties.

Biochar and ash show a significant effect on soil pH in the alkaline direction. Biomass ash is strongly alkaline—even when used in moderate doses, it noticeably raises the pH of the soil and neutralizes its acidity [31]. Biochar (especially obtained at high temperatures) is also most often alkaline and has a deacidifying effect, albeit weaker than ash. This is due to the presence in biochar of ash rich in Ca, K and Mg; studies in tea plantations confirm that the addition of biochar can raise the pH of acidic soil from ~ 4.5 to ~ 5.5 (the effect depends on the dose and type of biochar) [36]. The opposite trend is observed with long-term use of only nitrogen mineral fertilizers—it leads to soil acidification. Intensive fertilization with nitrogen in the form of ammonium accelerates the acidification process and, in addition, the use of physiologically acidic salts such as superphosphate or potassium chloride can exacerbate the decline in pH [38]. Organic fertilizers counteract these negative phenomena—they bring calcium, magnesium compounds and other buffers to the soil, which buffer the reaction. Manure often has a slightly alkaline reaction, so it helps maintain a pH closer to neutral. Combined application of mineral and organic fertilizers usually stabilizes the pH. In field experiments, combining manure with mineral fertilizer raised soil pH more than ammonium nitrate alone [34].

Organic materials (manure, compost, biochar) play a key role in improving soil structure and its water-holding capacity. NPK mineral fertilizers, providing only soluble salts, have no beneficial effect on the physical properties of the soil, improving neither soil aggregation, water holding capacity nor humus content [33]. In turn, regular fertilization with manure or compost clearly increases the level of organic matter, which translates into a better tubular structure, higher porosity and water holding capacity of the soil. Humus acts like a sponge—it retains water and minerals, reducing vulnerability to drought and leaching. Similar, though long-lasting, effects are produced by biochar: its highly porous structure increases the retention of water and plant-available ions in the soil [36]. Studies have noted that biochar additives improve soil water storage and nutrient availability, contributing to better plant growth [36]. Also, organic fertilizers improve physical properties—for example, they increase water holding capacity and soil permeability and aeration by loosening the soil profile and stimulating biological life [34]. Biomass ash, which lacks an organic fraction, has a limited effect on soil structure—it can only slightly improve the properties of compacted soils by contributing calcium and by neutralizing excessive acidity. It does not increase the soil's ability to store water the way organic additives do.

Another aspect is the effect on the soil microflora. Fertile soil requires active microorganisms, and these in turn need a source of organic carbon and the right environmental conditions. Mineral fertilizers by themselves do not provide nutrients for microflora—their application (especially without the addition of organic matter) usually does not result in a significant increase in the abundance of soil microorganisms [33]. In the long term, exclusive mineral chemistry may even reduce microbial biodiversity due to acidification and lack of fresh matter for heterotrophs. Organic fertilizers have the opposite effect: they introduce readily available carbon and energy compounds into the soil, stimulating microbial growth. Soils fertilized with compost or manure generally show higher enzymatic activity and greater microbial biomass than soils under exclusive NPK [36]. Biochar influences the microflora indirectly—although it is not an easily decomposed food (it is recalcitrant carbon), but due to its porosity it creates ecological niches conducive to colonization by microorganisms and improves conditions (pH, humidity) in the root zone [36]. Many studies have reported an increase in microbial biomass after the addition of biochar, especially when combined with nitrogen fertilization, which is associated with improved nitrogen retention in the soil and reduced toxic ammonia peaks [37]. Ash, by raising the pH of acidic soils, can stimulate microbial activity—deacidification and an influx of readily available macronutrients promotes soil microorganisms [31]. Demeyer et al. noted that the application of ash in moderate doses improves the chemical and physical conditions of the soil, which translates into an increase in the rate of mineralization of organic matter and the abundance of microorganisms compared to non-limed soil [39]. Caution should be exercised with very high doses of ash—raising the pH and salinity too much can selectively affect the microflora, favoring alkaliphilic bacteria, for example, and inhibiting other groups of organisms [40]. The addition of organic matter (compost, manure, biochar) creates better conditions for soil life than the use of mineral salts alone.

Mineral NPK fertilizers have the highest immediate yield-forming efficiency—they can significantly increase yields in a short period of time due to their high concentration of readily available forms of components. With balanced mineral fertilization, the maximum potential yield of a crop in a given season is usually achieved. In the long term, maintaining high soil productivity requires attention to soil fertility (structure, humus, microflora), which mineral salts alone do not provide. Here, organic fertilizers and soil improvers gain the upper hand, which, in addition to the nutrients themselves, improve the soil's ability to bear crops in subsequent years. In field experiments, yields obtained with balanced organic fertilization can match those on mineral fertilization, especially when sufficient rates are applied and the soil gradually improves in quality (increase in humus content, better retention of nutrients) [33]. The most favorable results are usually achieved with a combination of mineral and organic fertilization. Integrated strategies provide both fast-acting mineral ions and long-term reserves from organic matter—so plants don't suffer deficiencies at critical stages, and the soil gains fertility. For example, in a 3-year orchard experiment, a combination of a reduced rate of mineral fertilizers (60% of the standard rate) with the addition of organic substrate and sheep manure significantly increased soil fertility and yield compared to purely mineral fertilization [34]. This mixed strategy provided a full set of macronutrients, combining the immediate effect of manure (rich in rapidly available nitrogen) with slow-acting organic material, so that plant nutrition was evenly distributed throughout the growing season [34]. At the same time, there was an improvement in yield quality (better coloration and sugar content of the fruit) and an increase in soil biological activity under the influence of such combined fertilization [34]. Also in agricultural field crops, combining fertilizers—for example, the application of manure and mineral fertilizers at the same time—results in higher yields than each of them separately, due to synergistic action (the improvement of soil properties by organic matter increases the efficiency of mineral utilization by plants). Biochar, applied alone, usually does not result in a significant yield increase, but its value is revealed when combined with mineral or natural fertilizers. Meta-analyses of studies have shown that the addition of biochar together with fertilizers increases fertilizer efficiency and translates into higher yields than with conventional fertilization. For example, application of biochar together with mineral fertilizers increases yields compared to mineral fertilization alone [37]. On fertile soils, the impact will be less, and biochar may not give a significant yield increase without additional nitrogen fertilization. The yield-forming efficiency of integrated strategies (combining ash or biochar with traditional fertilizers) is very promising from the perspective of sustainable agriculture. Biomass ash can replace a significant portion of phosphorus and potassium fertilizers—research on P-rich plant ash has shown that it matches superphosphate in terms of yield and plant phosphorus content, although it has a slightly slower effect [31]. At the same time, ash provides calcium and micronutrients that are lacking in typical NPK fertilizers, which sometimes translates into healthier plants (fewer symptoms of micronutrient deficiencies) and better crop quality. Ash does not contain nitrogen—so in practice it cannot fully replace nitrogen fertilizers and must be used in conjunction with them or with nitrogen-fixing crops. Traditional mineral fertilizers remain the most effective in increasing yields quickly, but the best long-term results are achieved by supplementing them with organic matter (manure, compost) and enhancers such as biochar or ash. Such a holistic fertilization program makes it possible to simultaneously meet the nutritional needs of crops and improve soil properties, translating into high and stable productivity of the agroecosystem [34].

References

1. S.V. Vassilev, C.G. Vassileva, D. Baxter, Trace element concentrations and associations in some biomass ashes. Fuel **129**, 292–313 (2014). https://doi.org/10.1016/J.FUEL.2014.04.001
2. Z. Romanowska-Duda, R. Janas, M. Grzesik, Increasing fertilization efficiency of biomass ash by the synergistically acting digestate and extract from water plants sequestering CO_2 in sorghum crops. Molecules **29**(18), 4397 (2024). https://doi.org/10.3390/MOLECULES29184397
3. W. Buss, S. Jansson, O. Mašek, Unexplored potential of novel biochar-ash composites for use as organo-mineral fertilizers. J. Clean. Prod. **208**, 960–967 (2019). https://doi.org/10.1016/J.JCLEPRO.2018.10.189
4. J.A. Antonangelo, X. Sun, H.D.J. Eufrade-Junior, Biochar impact on soil health and tree-based crops: a review. Biochar **7**(1), 1–28 (2025). https://doi.org/10.1007/S42773-025-00450-6
5. X. Li et al., A global dataset of biochar application effects on crop yield, soil properties, and greenhouse gas emissions. Sci. Data **11**(1), 57 (2024). https://doi.org/10.1038/S41597-023-02867-9
6. C. Cruz-Paredes, Á. López-García, G.H. Rubæk, M.F. Hovmand, P. Sørensen, R. Kjøller, Risk assessment of replacing conventional P fertilizers with biomass ash: residual effects on plant yield, nutrition, cadmium accumulation and mycorrhizal status. Sci. Total Environ. **575**, 1168–1176 (2017). https://doi.org/10.1016/j.scitotenv.2016.09.194
7. A. Buntić et al., Synergistic effect of Bacillus isolates and biomass ash on soil and plant quality: a preliminary pot-experiment with the analysis of potentially toxic elements. Zemljiste Biljka **70**(2), 42–55 (2021). https://doi.org/10.5937/ZemBilj2102042B
8. M. Czernicka, C. Puchalski, R. Pawlak, M. Szostek, E. Szpunar-Krok, Analysis of the free amino acid profile of barley grain from organic fertilisation with ash from biomass combustion. Molecules **29**(1), 95 (2023). https://doi.org/10.3390/MOLECULES29010095
9. I. Singh, H.S. Thind, S. Sharma, Y. Singh, M. Amin Bhat, Phosphorus fertilizing potential of biomass ashes and their effect on bioavailability of micronutrients in wheat (*Triticum aestivum* L.). J. Appl. Nat. Sci. **9**(2), 744–753 (2017). Accessed 12 June 2025. [Online]. Available: www.jans.ansfoundation.org
10. L. Ye, M. Camps-Arbestain, Q. Shen, J. Lehmann, B. Singh, M. Sabir, Biochar effects on crop yields with and without fertilizer: a meta-analysis of field studies using separate controls. Soil Use Manag. **36**(1), 2–18 (2020). https://doi.org/10.1111/SUM.12546
11. S.C. Thomas, N. Gale, Biochar and forest restoration: a review and meta-analysis of tree growth responses. New For. (Dordr.) **46**(5–6), 931–946 (2015). https://doi.org/10.1007/S11056-015-9491-7/FIGURES/2
12. E. Szpunar-Krok, M. Szostek, R. Pawlak, J. Gorzelany, D. Migut, Effect of fertilisation with ash from biomass combustion on the mechanical properties of potato tubers (*Solanum tuberosum* L.) grown in two types of soil. Agronomy **12**(2), 379 (2022). https://doi.org/10.3390/AGRONOMY12020379/S1
13. K. Pycia, E. Szupnar-Krok, M. Szostek, R. Pawlak, L. Juszczak, Effect of soil type and application of ecological fertilizer composed of ash from biomass combustion on selected physicochemical, thermal, and rheological properties of potato starch. Molecules **27**(13), 4318 (2022). https://doi.org/10.3390/MOLECULES27134318
14. B. Saletnik, C. Puchalski, Suitability of biochar and biomass ash in basket willow (*Salix viminalis* L.) cultivation. Agronomy **9**(10), 577 (2019). https://doi.org/10.3390/AGRONOMY9100577

15. M.O.A. Mollick, A.K. Paul, I. Alam, M.M. Sumon, Effect of biochar on yield and quality of potato (*Solanum tuberosum*) tuber. Int. J. Bio-resour. Stress Manag. **11**(5), 445–450 (2020). https://doi.org/10.23910/1.2020.2140
16. R. Edussuriya, A.U. Rajapaksha, C. Jayasinghe, C. Pathirana, M. Vithanage, Influence of biochar on growth performances, yield of root and tuber crops and controlling plant-parasitic nematodes. Biochar **5**(1), 1–15 (2023). https://doi.org/10.1007/S42773-023-00261-7/FIGURES/1
17. T.M. Agbede et al., Impacts of poultry manure and biochar amendments on the nutrients in sweet potato leaves and the minerals in the storage roots. Sci. Rep. **14**(1), 1–15 (2024). https://doi.org/10.1038/s41598-024-67486-9
18. P. Tammeorg, A. Simojoki, P. Mäkelä, F.L. Stoddard, L. Alakukku, J. Helenius, Short-term effects of biochar on soil properties and wheat yield formation with meat bone meal and inorganic fertiliser on a boreal loamy sand. Agric. Ecosyst. Environ. **191**, 108–116 (2014). https://doi.org/10.1016/J.AGEE.2014.01.007
19. L. Du et al., Heterogeneous impact of soil acidification on crop yield reduction and its regulatory variables: a global meta-analysis. Field Crops Res. **319**, 109643 (2024). https://doi.org/10.1016/J.FCR.2024.109643
20. J. Wierzbowska, S. Sienkiewicz, P. Zarczyński, S. Krzebietke, Environmental application of ash from incinerated biomass. Agronomy **10**(4), 482 (2020). https://doi.org/10.3390/AGRONOMY10040482
21. S.K. Pradhan, J.K. Holopainen, H. Heinonen-Tanski, Stored human urine supplemented with wood ash as fertilizer in tomato (*Solanum lycopersicum*) cultivation and its impacts on fruit yield and quality. J. Agric. Food Chem. **57**(16), 7612–7617 (2009). https://doi.org/10.1021/JF9018917
22. E. Ibukunoluwa Moyin-Jesu, Use of different organic fertilizers on soil fertility improvement, growth and head yield parameters of cabbage (*Brassica oleraceae* L.). Int. J. Recycl. Org. Waste Agric. **4**(4), 291–298 (2015). https://doi.org/10.1007/S40093-015-0108-0/TABLES/6
23. G. Sahu, A. Ghosh Bag, N. Chatterjee, A.K. Mukherjee, Potential use of flyash in agriculture: a way to improve soil health. J. Pharmacogn. Phytochem. **6**(6), 873–880 (2017). Accessed 12 June 2025. [Online]. Available: https://www.phytojournal.com/archives/2017.v6.i6.2172/potential-use-of-flyash-in-agriculture-a-way-to-improve-soil-health
24. C.O. Rusănescu, M. Rusănescu, Application of fly ash obtained from the incineration of municipal solid waste in agriculture. Appl. Sci. **13**(5), 3246 (2023). https://doi.org/10.3390/APP13053246
25. D. Jabborova et al., Beneficial features of biochar and arbuscular mycorrhiza for improving spinach plant growth, root morphological traits, physiological properties, and soil enzymatic activities. J. Fungi **7**(7), 571 (2021). https://doi.org/10.3390/JOF7070571
26. P. Ofori, G. Asamoah, B. Amoah, K.O.A. Agyeman, E. Yeboah, Combined application of poultry litter biochar and NPK fertilizer improves cabbage yield and soil chemical properties. Open Agric. **6**(1), 356–368 (2021). https://doi.org/10.1515/OPAG-2021-0217
27. A. Obadi et al., Enhancement in tomato yield and quality using biochar amendments in greenhouse under salinity and drought stress. Plants **13**(12), 1634 (2024). https://doi.org/10.3390/PLANTS13121634
28. G.H.D. Agbna, S. Dongli, L. Zhipeng, N.A. Elshaikh, S. Guangcheng, L.C. Timm, Effects of deficit irrigation and biochar addition on the growth, yield, and quality of tomato. Sci. Hortic. **222**, 90–101 (2017). https://doi.org/10.1016/J.SCIENTA.2017.05.004
29. W. Li, J. Gao, S. Zhou, F. Zhou, Effect of biochar on apple yield and quality in aged apple orchards on the Loess Plateau (China). Agronomy **14**(6), 1125 (2024). https://doi.org/10.3390/AGRONOMY14061125/S1
30. V. Vijay et al., Review of large-scale biochar field-trials for soil amendment and the observed influences on crop yield variations. Front. Energy Res. **9**, 710766 (2021). https://doi.org/10.3389/FENRG.2021.710766/XML/NLM
31. K. Schiemenz, B. Eichler-Löbermann, Biomass ashes and their phosphorus fertilizing effect on different crops. Nutr. Cycl. Agroecosyst. **87**(3), 471–482 (2010). https://doi.org/10.1007/S10705-010-9353-9/TABLES/9

32. C. Jian et al., Effects of wood ash fertilizer on element dynamics in soil solution and crop uptake. Agronomy **15**(5), 1097 (2025). https://doi.org/10.3390/AGRONOMY15051097
33. L.C. Dincă, P. Grenni, C. Onet, A. Onet, Fertilization and soil microbial community: a review. Appl. Sci. **12**(3), 1198 (2022). https://doi.org/10.3390/APP12031198
34. L. Zhuang et al., A comprehensive study on the impact of chemical fertilizer reduction and organic manure application on soil fertility and apple orchard productivity. Agronomy **14**(7), 1398 (2024). https://doi.org/10.3390/AGRONOMY14071398
35. S.K. Shamila, S.S. Udumann, N.S. Dissanayaka, K. Rajaratnam, A.J. Atapattu, Assessing the impact of king coconut husk ash and biochar, combined with chemical fertilizer application, on enhancing soil fertility in coconut plantations. Crops **4**(2), 227–241 (2024). https://doi.org/10.3390/CROPS4020017/S1
36. R. Yin et al., Biochar, organic fertilizer, and bio-organic fertilizer improve soil fertility and tea quality. Agronomy **14**(10), 2339 (2024). https://doi.org/10.3390/AGRONOMY14102339
37. L. Zhang et al., Meta-analysis of the response of the productivity of different crops to parameters and processes in soil nitrogen cycle under biochar addition. Agronomy **12**(8), 1857 (2022). https://doi.org/10.3390/AGRONOMY12081857/S1
38. P. Tkaczyk, A. Mocek-Płóciniak, M. Skowrońska, W. Bednarek, S. Kuśmierz, E. Zawierucha, The mineral fertilizer-dependent chemical parameters of soil acidification under field conditions. Sustainability **12**(17), 7165 (2020). https://doi.org/10.3390/SU12177165
39. A. Demeyer, J.C. Voundi Nkana, M.G. Verloo, Characteristics of wood ash and influence on soil properties and nutrient uptake: an overview. Bioresour. Technol. **77**(3), 287–295 (2001). https://doi.org/10.1016/S0960-8524(00)00043-2
40. T. Bang-Andreasen et al., Wood ash induced pH changes strongly affect soil bacterial numbers and community composition. Front. Microbiol. **8**(JUL), 1400 (2017). https://doi.org/10.3389/FMICB.2017.01400/FULL

Chapter 6
Effects of Biomass Ash and Biochars on Soil Properties

6.1 Introduction

Biomass used for energy purposes generates two important byproducts: the ash produced after its combustion and the biochar obtained through pyrolysis. Both of these materials are used as soil additives, attracting much interest in the last decade. Biomass (e.g., wood) ash is increasingly treated as a substitute for lime and mineral fertilizer—when added to soil, it reduces its acidification and enriches it with nutrients [1]. Biochar, on the other hand, is defined as a highly carbonized, porous material formed from biomass under anaerobic conditions that has a high content of stable carbon. Research indicates that biochar is a multifunctional, sustainable soil improver that increases soil fertility, captures contaminants and contributes to improved plant growth [2]. Also important is the long-term fixation of carbon in the soil after biochar application, which promotes carbon sequestration and climate change mitigation [3]. Both ash and biochar introduced into the soil have a complex effect on a number of physical, chemical, and biological properties of that soil.

6.2 Volumetric Density and Porosity

The addition of both biomass ash and biochar most often leads to a reduction in the bulk density (BD) of the soil and an increase in its porosity. This has beneficial effects on structure—the soil becomes more porous. Field studies have shown that the application of biomass ash at doses of 100–500 kg K_2O/ha significantly reduced soil density in the arable layer (e.g., by about 0.12–0.15 Mg/m^3 in the 0–5 cm layer at the highest dose of 500 kg/ha). At the same time, there was a decrease in penetrometry resistance (PR) of the soil at higher ash doses, indicating a reduction in compactness and facilitation of root growth [4]. Similar effects are found after the application of biochar: a global meta-analysis of 59 studies found an average decrease in soil

K. Chojnacka et al., *Sustainable Fertilizers: Utilizing Biomass Ashes and Biochars*, SpringerBriefs in Applied Sciences and Technology,
https://doi.org/10.1007/978-3-032-03765-7_6

density of up to 29% and an increase in total porosity of 59% after the addition of biochar [3]. The reduction in density is due to the very low bulk density of biochar and its porous structure—the biochar particles act as light soil space fillers. Wood ash, on the other hand, although composed of fine mineral particles, often contains significant amounts of calcium and other cations that promote the flocculation of soil particles, which improves the formation of a porous aggregate structure instead of plugging the pores [4]. As a result, both additives reduce the phenomenon of over-compaction of soil as indicated, among other things, by the observation that biomass ash contributes to lower BD, which can have a beneficial effect on soil microorganisms through improved aeration [5].

6.3 Water Content and Water Retention

Both biochar and biomass ash affect the water management of soils, usually improving the soil's ability to retain water. Biomass ash has hygroscopic properties and provides cations (Ca^{2+}, K^{+}, and Mg^{2+}) that promote the formation of a tubular structure, resulting in better water retention in soil profiles. In a two-year field experiment on podzolic soil, it was found that already the lowest ash application rate (100 kg/ha in terms of K_2O) increased the field moisture content of the soil in all tested layers, while the highest rate (500 kg/ha) raised the volumetric moisture content of the 30–35 cm layer by about 3% compared to the control [4]. At the same time, the ash reduced the content of easily dispersible clay particles, reducing the susceptibility of the soil surface to crusting [4]. This is important because a lower propensity for crusting translates into better rainwater absorption and reduced surface runoff and erosion. Biochar, on the other hand, acts as a sponge in the soil due to its highly porous structure (micro- and mesoscopic pores), increasing water holding capacity and water availability for plants [6]. The pores of biochar can store water while, unlike the very fine pores of clay, making it available to roots at reduced water potential. Consequently, sandy soils and soils low in organic matter particularly benefit from water retention when biochar is added [3, 7]. For example, in a subtropical soil, an increase in maximum water holding capacity and field moisture was found, resulting in improved water-use efficiency of plants [2]. It is worth noting that very high doses of biochar can sometimes reduce water conductivity due to excessive pore filling with fine carbon or its hydrophobicity just after pyrolysis [8, 9].

6.4 Soil Structure and Aggregate Stability

The introduction of biochar and ash has beneficial effects on the structural properties of the soil, including the formation and stability of soil aggregates. Wood ash provides calcium and potassium, which act as flocculating agents for clay particles, promoting the formation of larger soil aggregates instead of individual, dispersed grains. As mentioned, studies have shown a reduction in the proportion of easily disintegrated clay fraction after the application of ash [4], meaning that soil aggregates became more resistant to disintegration in water. In other words, the stability of the structure has been improved, which translates into a lower tendency for the soil to become sodden and eroded. Biochar, on the other hand, acts as a scaffold for aggregates, and its porous particles with a large internal surface area can be encapsulated by clay-humus complexes, and microorganisms that produce binder substances (e.g., mycelium, bacterial polysaccharides) populate the pores of biochar, strengthening the bonds between soil particles [2]. Studies confirm that the addition of biochar increases the wet granular (water-resistant) proportion of aggregates according to a literature review, the stability of wet aggregates increased on average from several to tens of percent after biochar, depending on the soil [2]. Although in some soils with very high humus content the effect was sometimes less pronounced, the overall trend indicates improved aggregation due to biochar. However, it is worth noting that there are conflicting reports. One more recent study showed a decrease in the stability of aggregates in the soil after a very high dose of pure biochar, which the authors explained by the disruption of natural connections and leaching of some of the ash from the biochar during intensive irrigation [2]. Despite such exceptions, most results confirm the beneficial effect of biochar on structure [10], especially when it is used in combination with organic matter (e.g., compost), which further increases aggregate resistance and reduces soil losses.

6.5 Water Infiltration and Permeability

Improved soil porosity and structure translates directly into better hydrodynamic parameters, such as water infiltration rates and soil hydraulic conductivity. Studies show that soils enriched with ash or biochar absorb rainwater faster and allow it to percolate deeper [11, 12]. In a field trial in India, it was found that a dose of 2 t/ha of biochar increased the infiltration rate by several-odd percent and raised the saturated hydraulic conductivity in the crop layer [2]. When biochar was additionally combined with organic fertilizers, the effect was even more pronounced—infiltration increased by a total of 4–38% and water permeability by 0.9–13.7% relative to compost alone [2]. Biomass ash, due to the reduction of surface crusting and increased macroporosity, also improves water infiltration. Lower soil compaction and more stable aggregates in the plow layer result in higher infiltration and lower surface runoff during heavy rainfall [4]. In addition, improving the structure of heavy soils by ash can increase the proportion of drainage pores, which promotes the drainage of excess water and prevents stagnation. In general, both types of additives improve the water balance of soils: during periods of drought, the soil stores water better and retains moisture thanks to biochar and ash, while during periods of excess rainfall, the soil better transmits it, protecting crops from flooding. These beneficial changes in physical properties translate into reduced soil degradation (less erosion, compaction) and create favorable conditions for the growth of the plant root system [13, 14].

6.6 Soil pH

Both biomass ash and biochar are among the highly alkaline soil additives, meaning that they are effective in raising soil pH [15]. Wood ash generally has a pH in the range of 10–13 (for example, the ash used in experiments in Poland had a pH of ~ 12.8) [16] and contains carbonates, oxides and hydroxides of alkaline and basic metals. As a result, it has a deacidifying effect similar to lime, neutralizing H^+ ions in the soil solution. In an acidic podzolic soil, a single dose of biomass ash increased the pH from about 4.1 to 5.77 in the arable layer [16]. However, even a small increase in pH can significantly affect soil life—it has been observed that ash at doses as low as 100–200 kg/ha induces a statistically significant increase in pH in the topsoil layer, beneficial to most microorganisms and plants [15]. Biochar, on the other hand, has a pH that depends on the pyrolysis temperature and the starting material. A meta-analysis indicates that overall, biochar raises soil pH [3]. The mechanism is similar to that of ash, biochar contains ash (undecomposed biomass minerals), often rich in Ca, K, and Mg, which is released into the soil and binds protons, abolishing acidity [17, 18]. In addition, the surface of biochar is oxidized over time, generating additional basic (carboxyl) functional groups [19]. In sum, the addition of the materials in question produces a permanent calcifying effect in the soil, which is particularly beneficial in environments with excessive acidity (e.g. forest soils, soils after intensive nitrogen fertilization). Numerous studies confirm that wood ash can be effectively used to deacidify acidic soils, often acting as effectively as traditional lime [20]. It should be noted, however, that the pH-raising effect depends on the application rate. Moderate amounts of ash improve the pH to an optimal level, while excessive doses can exceed alkalinity. For example, in a forestry experiment, lower doses of ash (5–10 Mg/ha) favorably neutralized acidic soil, improving tree growth, but the highest dose of 20 Mg/ha caused an excessive increase in pH and initially negatively affected the condition of seedlings [21]. Therefore, it is recommended to apply ash in reasonable amounts, preceded by a soil buffer analysis.

6.7 Sorption Capacity and Organic Matter

The addition of biochar clearly increases the content of stable organic matter (organic carbon) in the soil, while ash, lacking carbon, affects organic matter indirectly. Biochar consists of 60–90% elemental carbon in an aromatic form that resists decomposition, making it a stable carbon accumulator in the soil. On average, a biochar application increases total soil organic carbon (SOC) by about 27% [3], with this increase sustained over many years. The carbon of biochar has an average soil life of hundreds of years, and much of it can remain in the soil for up to millennia [22, 23]. An increase in the stock of stable organic matter translates into a number of beneficial chemical effects: cation exchange capacity (CEC) increases, aggregate structure improves, and the ability to buffer pH and retain water increases. Meta-analyses indicate that biochar increases soil CEC by about 10–20% on average, and in sandy and humus-poor soils the increase is sometimes even greater [3]. This is because the surface of the biochar over time becomes covered with functional groups (carboxylic, phenolic) that confer a negative charge, and in addition, the biochar itself often already contains some basal CEC associated with the ash present and the porous structure. As a result, the soil's ability to sorb and exchange nutrient cations increases, which promotes their retention and availability to plants [24, 25]. Biomass ash itself does not have CEC, but it affects the soil sorption complex by saturating it with cations and raising the pH. In acidic soils, much of the sorption complex (especially with humus and Fe/Al oxides) is pH-dependent. An increase in pH due to ash reveals additional negative charges on the surface of organic and mineral particles. Thus, ash indirectly increases the effective sorption capacity of acidic soil, and this is expressed by an increase in the saturation of basic resources and the sum of basic cations on the sorption complex [26]. Moreover, by providing Ca^{2+}, the ash displaces H^+ and Al^{3+} ions from the sorption complex, replacing them with base cations. The soil becomes more saturated with calcium and magnesium, which stabilizes the pH in the neutral range and improves fertility. As for organic matter, ash, while not contributing its own carbon, can initially have a stimulating effect on the mineralization of existing humus [27]. Shortly after the ash application, an increase in the rate of soil respiration and increased mineralization of organic carbon was observed [27]. However, this does not imply a permanent loss of humus. On a longer time scale, improved conditions and increased plant biomass production may result in a greater influx of plant residues into the soil and the restoration of organic matter reserves. One incubation study found that wood ash alone made little difference in soil organic carbon content, while when combined with straw, it contributed to greater carbon sequestration (both organic and inorganic carbonate) than straw alone [28]. This suggests that ash may stabilize additional carbon from external sources (e.g., by forming calcium carbonates from CO_2 of the soil). However, the role of ash in the organic matter balance is mainly to regulate soil conditions.

6.8 Nitrogen and Biogeochemical Cycles

A specific issue is the effect of ash and biochar on nitrogen metabolism in the soil. Ash, while not containing nitrogen, does not supply it directly, but by raising the pH and adding lime it stimulates the nitrification of ammonium in the soil [29, 30]. Deacidified and calcified soil promotes nitrifying bacteria that convert ammonium to nitrate. At the same time, due to improved nutrient conditions, overall microbial activity increases, which can result in faster turnover of organic matter and release of nitrogen from organic matter [27]. However, excessive increases in nitrification can also increase nitrogen losses (NO_3^- leaching, N_2O emissions)—here it can be helpful to combine ash with moderate nitrogen fertilization to ensure plants capture the generated nitrates [21]. Biochar, on the other hand, often has the opposite effect, reducing nitrogen losses, although it provides little nitrogen itself (unless it comes from fertilized sludge or protein-rich biomass). The action of biochar is to adsorb ammonium ions on its surface (thanks to CEC), preventing them from leaching out of the profile, and to create more favorable aerobic conditions in the soil, which promotes the full denitrification of nitrates to N_2 instead of nitrous oxide (N_2O). Meta-analyses confirm that the addition of biochar reduces nitrous oxide emissions from soil—on average, about 10–15% reduction in N_2O flux is observed [31]. In soil with biochar, an increase of ~ 40% in nitrification potential was observed with a ~ 12% decrease in N_2O emissions, indicating more complete denitrification processes [31]. In other words, biochar helps retain nitrogen in the soil–plant system while giving it back to plants as needed. However, these effects are variable—depending on the type of soil and biochar. In sandy, poor soils, biochar clearly improves nitrogen availability to plants (by reducing leaching), while in fertile soils it can somewhat adsorb available nitrogen and temporarily reduce the amount of nitrogen for crops until desorption or mineralization of the extra matter occurs [2]. Therefore, it is sometimes observed that biochar must be applied simultaneously with nitrogen fertilizer or compost to ensure nitrogen abundance and optimal use of biochar's sorption properties [2].

6.9 Pollutants and Heavy Metals

An important chemical aspect of biomass ash use is potential heavy metal contamination. Biomass during growth takes up not only nutrients from the ground, but also metals such as cadmium, lead, mercury, and arsenic. During the combustion process, these metals are not degraded and concentrate in the ash [32]. Therefore, ash must meet safety standards before agricultural use, this is especially true for ash from sewage sludge [33]. Fortunately, analyses indicate that most biomass ash has heavy metal contents within acceptable ranges and, at moderate doses, does not cause soil limits to be exceeded [34, 35]. The pH-raising effect of ash has a mitigating effect on the mobility of many heavy metals. In deacidified soil, most of them precipitate into less bioavailable forms. Thus, paradoxically, ash can reduce the bioavailability of endogenous heavy metals in acidic soils, reducing their uptake by plants, even if it introduces some amounts. The opposite effect applies only to sodium. Ashes with high Na content can increase salinity and sodium leaching, potentially worsening the structure. Biochar, on the other hand, is known for its strong sorptive properties against heavy metals. Due to its large surface area and the presence of various functional groups, biochar binds metal cations by complexation, ion exchange, electrostatic adsorption, or reduction to less mobile forms [36, 37]. In a number of studies on contaminated soils, the addition of biochar has been shown to immobilize metals such as cadmium, zinc, lead, and nickel, reducing their concentration in the soil solution and uptake by plants [38]. For example, in a pot experiment with perennial ryegrass, biochar modified with MnO_2 nanoparticles significantly reduced the uptake of Pb, Cd, Zn, and Ni by plants, significantly improving their growth [38]. Even unmodified biochar typically reduces metal availability by raising pH and sorption, although its effectiveness is sometimes lower than that of activated biochar [39–41]. In conclusion, biochar promotes the stabilization of metallic contaminants in the soil, making the reclaimed ground safer for plants and groundwater. It is worth mentioning that biochar adsorbs not only metals, but also organic contaminants, which may be desirable in phytoremediation [42].

6.10 Microbial Structure and Diversity

Additives such as biochar and ash can alter the species composition of soil microflora. Biochar often promotes the growth of aerobic microbial groups, especially nitrogen-converting bacteria (nitrifiers, nitrogen assimilating bacteria) and symbiotic fungi [43]. A meta-analysis showed that biochar stimulates the activity of nitrification genes and nitrogen cycle enzymes (such as urease, nitrate reductase), with a slight decrease in the abundance of archaic nitrifiers (AOA) [31]. This could be interpreted to mean that conditions after biochar favor typical soil bacteria rather than extremophilic archaeon's, which prefer very acidic environments. As for species diversity, studies have yielded inconclusive results. An increase in diversity (e.g., Shannon index) of the microbiome is often observed after biochar, which is attributed to the creation of new microhabitats and a more balanced soil environment [3]. However, at very high doses of biochar or under conditions where the biochar contained certain contaminants (e.g., PAHs from insufficient pyrolysis), a decrease in diversity was observed, likely due to forced selection for strains tolerant of these conditions [44]. Moderate doses of biochar increase the taxonomic richness of the microflora, while excess can slightly decrease it [45]. In the case of biomass ash, new groups of microorganisms previously scarce in acidic soils often emerge [15]. Microbial revival results in an increase in the intensity of biological processes in the soil. Measures are the activities of soil enzymes, such as dehydrogenase (an indicator of microbial respiration activity), urease (the breakdown of urea to ammonium), phosphatase, cellulase, or glucosidase (the breakdown of complex sugars) [46]. In the presence of biochar, a strong increase in dehydrogenase activity was found, with a meta-analysis showing an average increase of 84% over control [31]. This is a very large change, suggesting more intensive decomposition of organic matter by microorganisms and greater electron flow in their metabolism. Similarly, the activities of other carbon cycle enzymes are increasing. Cellulase, which breaks down cellulose, increased activity by ~ 56% on average, invertase by ~ 21%. Interestingly, β-glucosidase itself did not change significantly or even decreased in some studies, which was linked to the fact that biochar sometimes adsorbs enzymes or reaction products. Nevertheless, overall, biochar clearly accelerates the soil carbon cycle—for example, a ~ 22% increase in CO_2 (soil respiration) has been reported following the addition of biochar [31]. This is not an undesirable phenomenon, as it mainly involves readily available compounds. The stable carbon pool of biochar remains in the soil, and the increased mineralization indicates a more intensive nutrient cycle. In the case of ash, an increase in enzymatic activity related to the nitrogen cycle is also noted. The addition of ash increased urease activity and the potential for nitrification, resulting in faster conversion of ammonium nitrogen to plant-available nitrate nitrogen [21]. In forest soils, the activity of alkaline phosphatase (an enzyme that releases phosphate from organic phosphorus) increased under the influence of ash, which is explained by an increase in pH and the need of plants for additional phosphorus under conditions of faster growth after liming with ash [16]. In general, the biological life of the soil activates: microorganisms decompose residues faster, more components circulate, the soil becomes more “alive”. It should be added, however, that certain enzymes can behave in the opposite way—for example, in one study, intensive growth of microflora next to biochar resulted in faster consumption of substrates and a decrease in the activity of cellulolytic enzymes, which the authors interpreted as a change in microbial strategy in the presence of more readily available carbon from biochar [31].

References

1. E. Błońska, W. Prażuch, P. Boroń, J. Lasota, Effects of wood ash on the soil properties and fungal community structure in a beech forest in Poland. Geoderma Reg. **34**, e00676 (2023). https://doi.org/10.1016/J.GEODRS.2023.E00676
2. P. Sharma et al., Effectiveness of biochar and compost on improving soil hydro-physical properties, crop yield and monetary returns in inceptisol subtropics. Saudi J. Biol. Sci. **28**(12), 7539 (2021). https://doi.org/10.1016/J.SJBS.2021.09.043
3. H. Singh, B.K. Northup, C.W. Rice, P.V.V. Prasad, Biochar applications influence soil physical and chemical properties, microbial diversity, and crop productivity: a meta-analysis. Biochar **4**(1), 1–17 (2022). https://doi.org/10.1007/S42773-022-00138-1/FIGURES/7
4. J. Czy et al., The impact of using different doses of biomass ash on some physical properties of podzolic soil under the cultivation of winter oilseed rape. Int. J. Environ. Res. Public Health **19**(11), 6693 (2022). https://doi.org/10.3390/IJERPH19116693
5. E.M. Bonfim-Silva, T. Martinez-Santos, T.J.A. da Silva, R.D. Rackel, E.A.R. Pinheiro, T.F. Duarte, Wood ash as a vegetative-growth promoter in soils with subsurface compaction. Rev. Bras. Eng. Agric. Ambient. **26**(4), 258–265 (2022). https://doi.org/10.1590/1807-1929/AGRIAMBI.V26N4P258-265
6. Z. Zhu, Y. Zhang, W. Tao, X. Zhang, Z. Xu, C. Xu, The biological effects of biochar on soil's physical and chemical characteristics: a review. Sustainability **17**(5), 2214 (2025). https://doi.org/10.3390/SU17052214
7. A.G. Alghamdi, A. Alkhasha, H.M. Ibrahim, Effect of biochar particle size on water retention and availability in a sandy loam soil. J. Saudi Chem. Soc. **24**(12), 1042–1050 (2020). https://doi.org/10.1016/j.jscs.2020.11.003
8. J. Zhao, T. Ren, Q. Zhang, Z. Du, Y. Wang, Effects of biochar amendment on soil thermal properties in the North China Plain. Soil Sci. Soc. Am. J. **80**(5), 1157–1166 (2016). https://doi.org/10.2136/SSSAJ2016.01.0020
9. Q. Yan et al., Effects of maize straw-derived biochar application on soil temperature, water conditions and growth of winter wheat. Eur. J. Soil Sci. **70**(6), 1280–1289 (2019). https://doi.org/10.1111/EJSS.12863

10. A. Khademalrasoul, N.J. Kuhn, L. Elsgaard, Y. Hu, B.V. Iversen, G. Heckrath, Short-term effects of biochar application on soil loss during a rainfall-runoff simulation. Soil Sci. **184**(1), 17–24 (2019). https://doi.org/10.1097/SS.0000000000000247
11. C. Mirko, S.A. Maria, T. Matteo, B. Emanuele, Spatial variability of soil physical and hydraulic properties in a durum wheat field: an assessment by the BEST-procedure. Water **11**(7), 1434 (2019). https://doi.org/10.3390/W11071434
12. W.D. Reynolds, C.F. Drury, C.S. Tan, C.A. Fox, X.M. Yang, Use of indicators and pore volume-function characteristics to quantify soil physical quality. Geoderma **152**(3–4), 252–263 (2009). https://doi.org/10.1016/J.GEODERMA.2009.06.009
13. H. Blanco-Canqui, Biochar and soil physical properties. Soil Sci. Soc. Am. J. **81**(4), 687–711 (2017). https://doi.org/10.2136/SSSAJ2017.01.0017
14. H. Ghazouani et al., Integrative effect of activated biochar to reduce water stress impact and enhance antioxidant capacity in crops. Sci. Total Environ. **905**, 166950 (2023). https://doi.org/10.1016/J.SCITOTENV.2023.166950
15. T. Bang-Andreasen et al., Wood ash induced pH changes strongly affect soil bacterial numbers and community composition. Front. Microbiol. **8**(JUL), 271559 (2017). https://doi.org/10.3389/FMICB.2017.01400/BIBTEX
16. M. Pastuszczak, J. Stanek-Tarkowska, M. Kačániová, Impact of soil fertilized with biomass ash on depth-related variability of culturable bacterial diversity and selected physicochemical parameters in spring barley cultivation. Int. J. Environ. Res. Public Health **19**(21), 13721 (2022). https://doi.org/10.3390/IJERPH192113721/S1
17. S. Gul, J.K. Whalen, B.W. Thomas, V. Sachdeva, H. Deng, Physico-chemical properties and microbial responses in biochar-amended soils: mechanisms and future directions. Agric. Ecosyst. Environ. **206**, 46–59 (2015). https://doi.org/10.1016/J.AGEE.2015.03.015
18. R. Chintala et al., Molecular characterization of biochars and their influence on microbiological properties of soil. J. Hazard. Mater. **279**, 244–256 (2014). https://doi.org/10.1016/J.JHAZMAT.2014.06.074
19. M. Qiu et al., Biochar for the removal of contaminants from soil and water: a review. Biochar **4**(1), 1–25 (2022). https://doi.org/10.1007/S42773-022-00146-1
20. R.M. Pitman, Wood ash use in forestry—a review of the environmental impacts. Forestry Int. J. For. Res. **79**(5), 563–588 (2006). https://doi.org/10.1093/FORESTRY/CPL041
21. J.Y. An, B.B. Park, Effects of wood ash and N fertilization on soil chemical properties and growth of *Zelkova serrata* across soil types. Sci. Rep. **11**(1), 1–13 (2021). https://doi.org/10.1038/s41598-021-93805-5
22. J. Lehmann, J. Gaunt, M. Rondon, Bio-char sequestration in terrestrial ecosystems—a review. Mitig. Adapt. Strateg. Glob. Change **11**(2), 403–427 (2006). https://doi.org/10.1007/S11027-005-9006-5/METRICS
23. D. Solomon et al., Molecular signature and sources of biochemical recalcitrance of organic C in Amazonian Dark Earths. Geochim. Cosmochim. Acta **71**(9), 2285–2298 (2007). https://doi.org/10.1016/J.GCA.2007.02.014
24. H. Blanco-Canqui, R. Lal, Crop residue removal impacts on soil productivity and environmental quality. CRC Crit. Rev. Plant Sci. **28**(3), 139–163 (2009). https://doi.org/10.1080/07352680902776507
25. A.K. Mensah, K.A. Frimpong, Biochar and/or compost applications improve soil properties, growth, and yield of maize grown in acidic rainforest and coastal savannah soils in Ghana. Int. J. Agron. **2018**(1), 6837404 (2018). https://doi.org/10.1155/2018/6837404
26. E. Manirakiza, B. Gagnon, N. Ziadi, Soil and plant cations as affected by application of wood ash, biochar, and papermill biosolids. Agron. J. **117**(1), 1 (2025). https://doi.org/10.1002/AGJ2.21714
27. R.L. Couch, N. Luckai, D. Morris, A. Diochon, Short-term effects of wood ash application on soil properties, growth, and foliar nutrition of picea mariana and picea glauca seedlings in a plantation trial. Can. J. Soil Sci. **101**(2), 203–215 (2021). https://doi.org/10.1139/CJSS-2020-0105/SUPPL_FILE/CJSS-2020-0105SUPPLA.DOCX

28. H. Zhao et al., Effect of straw and wood ash on soil carbon sequestration and bacterial community in a calcareous soil. Front. Microbiol. **13**, 926506 (2022). https://doi.org/10.3389/FMICB.2022.926506/BIBTEX
29. S. Zimmermann, B. Frey, Soil respiration and microbial properties in an acid forest soil: effects of wood ash. Soil Biol. Biochem. **34**(11), 1727–1737 (2002). https://doi.org/10.1016/S0038-0717(02)00160-8
30. A. Saarsalmi, E. Mälkönen, M. Kukkola, Effect of wood ash fertilization on soil chemical properties and stand nutrient status and growth of some coniferous stands in Finland. Scand. J. For. Res. **19**(3), 217–233 (2004). https://doi.org/10.1080/02827580410024124;JOURNAL:JOURNAL:SFOR20
31. P. Kerner et al., Microbial responses to biochar soil amendment and influential factors: a three-level meta-analysis. Environ. Sci. Technol. **57**(48), 19838–19848 (2023). https://doi.org/10.1021/ACS.EST.3C04201
32. A. Demirbaş, Heavy metal contents of fly ashes from selected biomass samples. Energy Sources **27**(13), 1269–1276 (2005). https://doi.org/10.1080/009083190519384
33. Y. You et al., Producing superphosphate with sewage sludge ash: assessment of phosphorus availability and potential toxic element contamination. Agronomy **11**(8), 1506 (2021). https://doi.org/10.3390/AGRONOMY11081506/S1
34. G. Zajac, J. Szyszlak-Bargłowicz, M. Szczepanik, Influence of biomass incineration temperature on the content of selected heavy metals in the ash used for fertilizing purposes. Appl. Sci. **9**(9), 1790 (2019). https://doi.org/10.3390/APP9091790
35. Y. Pei, M. Ike, K. Shiota, M. Takaoka, The impacts of furnace and fuel types on the hazardous heavy metal contents and leaching behavior of woody biomass fly ash. Fuel **372**, 132202 (2024). https://doi.org/10.1016/J.FUEL.2024.132202
36. H. Gong, L. Zhao, X. Rui, J. Hu, N. Zhu, A review of pristine and modified biochar immobilizing typical heavy metals in soil: applications and challenges. J. Hazard. Mater. **432**, 128668 (2022). https://doi.org/10.1016/J.JHAZMAT.2022.128668
37. C. Tu, J. Wei, F. Guan, Y. Liu, Y. Sun, Y. Luo, Biochar and bacteria inoculated biochar enhanced Cd and Cu immobilization and enzymatic activity in a polluted soil. Environ. Int. **137**, 105576 (2020). https://doi.org/10.1016/J.ENVINT.2020.105576
38. M.V. Ghandali, S. Safarzadeh, R. Ghasemi-Fasaei, S. Zeinali, Heavy metals immobilization and bioavailability in multi-metal contaminated soil under ryegrass cultivation as affected by ZnO and MnO_2 nanoparticle-modified biochar. Sci. Rep. **14**(1), 1–17 (2024). https://doi.org/10.1038/s41598-024-61270-5
39. H. Zhang et al., Efficient removal of heavy metal ions from wastewater and fixation of heavy metals in soil by manganese dioxide nanosorbents with tailored hollow mesoporous structure. Chem. Eng. J. **459**, 141583 (2023). https://doi.org/10.1016/J.CEJ.2023.141583
40. J. Fan et al., Remediation of cadmium and lead polluted soil using thiol-modified biochar. J. Hazard. Mater. **388**, 122037 (2020). https://doi.org/10.1016/J.JHAZMAT.2020.122037
41. Z. Meng, S. Huang, T. Xu, Z. Lin, J. Wu, Competitive adsorption, immobilization, and desorption risks of Cd, Ni, and Cu in saturated-unsaturated soils by biochar under combined aging. J. Hazard. Mater. **434**, 128903 (2022). https://doi.org/10.1016/J.JHAZMAT.2022.128903
42. F.U. Haider et al., Biochar application for remediation of organic toxic pollutants in contaminated soils; an update. Ecotoxicol. Environ. Saf. **248**, 114322 (2022). https://doi.org/10.1016/J.ECOENV.2022.114322
43. M. Kracmarova-Farren et al., Biochar-induced changes in soil microbial communities: a comparison of two feedstocks and pyrolysis temperatures. Environ. Microbiome **19**(1), 87 (2024). https://doi.org/10.1186/s40793-024-00631-z
44. J. Bai, B.R. De Almeida Moreira, Y. Bai, C.G. Nadar, Y. Feng, S. Yadav, Assessing biochar's impact on greenhouse gas emissions, microbial biomass, and enzyme activities in agricultural soils through meta-analysis and machine learning. Sci. Total Environ. **963**, 178541 (2025). https://doi.org/10.1016/J.SCITOTENV.2025.178541
45. X. Li, T. Wang, S.X. Chang, X. Jiang, Y. Song, Biochar increases soil microbial biomass but has variable effects on microbial diversity: a meta-analysis. Sci. Total Environ. **749**, 141593 (2020). https://doi.org/10.1016/j.scitotenv.2020.141593

46. X. Liao, H. Kang, G. Haidar, W. Wang, S. Malghani, The impact of biochar on the activities of soil nutrients acquisition enzymes is potentially controlled by the pyrolysis temperature: a meta-analysis. Geoderma **411**, 115692 (2022). https://doi.org/10.1016/j.geoderma.2021.115692

Chapter 7
Policy and Regulatory Framework

7.1 Introduction

The use of ashes from biomass combustion and biochar as fertilizers or soil improvers is gaining increasing attention in the context of a circular economy and sustainable agriculture. Biomass ash contains macronutrients (e.g., potassium, phosphorus, and calcium) and can increase soil pH, while biochar has high stability in soil and the ability to sequester carbon, helping to improve soil fertility and water retention. Despite these benefits, it is crucial that an appropriate legal framework and quality standards are in place to ensure the safe and effective use of these materials. Efforts have been made both internationally and in the European Union to regulate the status of biomass ash and biochar—from legal classification, to quality requirements, to their inclusion in the official fertilizer cycle [1, 2]. This chapter presents the current regulation of biomass ash and biochar-based fertilizers, discusses quality and safety standards for these products, and examines future agricultural policy perspectives in this area.

Regulations determine whether and under what conditions biomass ash and biochar can be used in agriculture. These materials are often produced as by-products or waste (e.g., ash from biomass CHP plants, biochar from pyrolysis of biomass waste) and are therefore subject to both waste and fertilizer regulations. The regulatory framework for biomass ash and biochar is discussed separately below.

7.2 Biomass Ashes as Fertilizers—Regulatory Framework

Globally and in Europe, the need to manage biomass combustion ashes in agriculture has been recognized, driven by the growing production of these ashes and the drive to recycle nutrients [1]. The EU's Thematic Strategy on Soil Protection had already emphasized the importance of enriching soils by recycling nutrients from waste.

K. Chojnacka et al., *Sustainable Fertilizers: Utilizing Biomass Ashes and Biochars*,
SpringerBriefs in Applied Sciences and Technology,
https://doi.org/10.1007/978-3-032-03765-7_7

Some EU member states have recognized the value of biomass ash as a source of nutrients and have implemented their own guidelines for its use, especially in forests (wood ash) or fields. In Scandinavian countries, for example, there are national regulations specifying permissible concentrations of heavy metals in ash used as fertilizer and minimum nutrient content to ensure the effectiveness of such fertilizer [2].

At the European Union level, there has long been a lack of a unified approach—biomass ash has been treated primarily as waste, which has hindered its direct agricultural use. Under the EU Waste Framework Directive 2008/98/EC, a material that is waste can cease to be considered waste only after it meets end-of-waste criteria or is considered a by-product [3]. For biomass ash, such criteria had to be clarified. In Poland, for example, pure plant ash (phyto ash) is currently classified as waste and may only be applied via an authorized R10 recovery permit under Poland's Environmental Protection Law. This means that a company wishing to use biomass ash on fields must obtain the appropriate permits for the recovery process. This legal status has inhibited wider implementation of the practice, despite evidence of the agricultural usefulness of ash (e.g., straw ash is rich in potassium and phosphorus) [4].

A breakthrough at the EU level came with the adoption of Regulation (EU) 2019/1009 on Fertilizing Products Regulation (FPR). The regulation, effective from July 2022, replaced previous legislation (Regulation (EC) No. 2003/2003) and opened the EU market to organic, waste and recycled fertilizers. Under the new law, a system of product functional categories (PFCs) and component material categories (CMCs) was introduced. This allows a fertilizer to be CE marked and sold throughout the EU if it meets the requirements for a functional category (e.g., fertilizer, soil improver, liming agent) and if all of its components fall within the permitted component material categories [5]. Biomass ash has been included in this system provided it meets certain criteria—Regulation 2019/1009 added CMC category 13: thermal oxidation materials or derivates for ash from biomass combustion. The criteria for CMC 13 (supplemented by the Commission's delegated acts) stipulate that the ash must come only from specific, clean biomass streams (e.g., clean wood, plant waste that does not exhibit hazardous properties) and meet contaminant limits before becoming an ingredient in a fertilizer product UE [5, 6]. If biomass ash meets the requirements of CMC 13, being derived solely from clean biomass and falling within the specified contaminant limits, it is no longer considered waste and may be marketed across the EU as a fertilizer. Under Regulation 2019/1009, CMC 13 allows clean-biomass ash to be marketed EU-wide once it meets feedstock purity and contaminant limits; compliant ash therefore exits 'waste' status and qualifies as a CE-marked fertilizer [2].

Regulation 2019/1009 opens an EU-wide CE route, yet Member States may still authorize non-CE ash fertilizers under national rules, so a domestic channel co-exists with the harmonized market-an arrangement that encourages producers to upgrade quality and obtain CE marking [3, 6].

7.3 Biochar as a Soil Improver—Regulatory Framework

Biochar, a carbonaceous material obtained through a thermochemical process (pyrolysis or gasification of biomass with limited oxygen), has also lived to see regulation, although somewhat later than ash. In many countries, biochar was initially treated as an experimental material, with no clearly defined status: product or waste [3]. In the United States and other countries, requirements for biochar are mainly set at the state or local level—for example, in the U.S., some states require registration of biochar as a fertilizer or soil ameliorant, and industry organizations have developed labeling guidelines for these [7]. There has been a lack of a uniform definition and quality standard, making it difficult to promote biochar on a wider scale.

Regulation (EU) 2019/1009 and its follow-up acts have formally opened the EU market to biochar [8]. To meet the mandate to assess biochar "without undue delay", the Commission issued Delegated Regulation (EU) 2021/2088, creating CMC 14—Pyrolysis and Gasification Materials and inserting biochar into Annex II of the Fertilizing Products Regulation [9]. CE-eligible biochar must: (i) originate solely from authorized feedstocks (plant residues, clean wood, pre-treated manure), with hazardous streams and untreated category 1 ABP excluded [10]; (ii) be produced under oxygen-limited conditions and show H/C_org < 0.7 to demonstrate carbon stability [8]; (iii) comply with contaminant ceilings of PAH16 ≤ 6 mg kg^{-1} DM and PCDD/F ≤ 20 ng WHO-TEQ kg^{-1} DM [8]; and (iv) contain $\leq 5\%$ additive mass, with extra limits on chlorine and thallium when additives are used [10, 11].

The introduction of biochar into EU fertilizer law was a breakthrough—it means that biochar meeting the requirements of CMC 14 can be a component of fertilizers, soil improvers or carbon binders, which will be CE marked and freely marketed in the EU market [12, 13]. As in the case of ash, member states can still allow biochar to be used on their territory under national regulations (e.g., as a soil improver not covered by Regulation 2019/1009), but EU certification becomes the gold standard, ensuring that the product has passed an assessment of compliance with strict safety standards [2].

It's worth mentioning that voluntary biochar quality certification systems, such as the European Biochar Certificate (EBC) and the International Biochar Initiative (IBI) guidelines, have emerged alongside regulations. These systems were in place before the EU regulated biochar and played an important role in preparing the ground for formal regulations. For example, the EBC certificate defines biochar use classes (agricultural, organic, technical, etc.) and permissible parameter ranges, which includes contaminant limits [14, 15]. EBC's requirements currently coincide with EU standards in key aspects—for example, the maximum total content of 16 PAHs in agricultural grade biochar has been set at 6 mg/kg, in line with planned EUeuropean-biochar.org limits. The EBC guideline likewise specifies maximum heavy metal concentrations, harmonizing them with national fertilizer limits (e.g., for ECB agricultural grade biochar, the limit for cadmium is 1.5 mg/kg, lead 150 mg/kg, zinc 400 mg/kg dry weight, etc.) [11, 14]. While these certifications are voluntary, they are often required by biochar consumers and supplement regulation—until the law was fully implemented, certification was sometimes proof that a product was safe.

For biomass ash and biochar, significant progress has been made in the regulatory framework. The European Union has incorporated these materials into the fertilizer law, establishing criteria to ensure that only safe and quality products reach the market. The implementation of these regulations in practice is only just beginning, and producers must adjust processes to meet the new requirements [1, 2]. Another key element is detailed quality and safety standards, discussed in the next section.

7.4 Quality and Safety Standards

The quality and safety of biomass ash and biochar for use in the soil environment are paramount—they determine the acceptability of these materials as fertilizers. Quality standards primarily include acceptable levels of contaminants (heavy metals, organic compounds, dioxins, etc.), but also parameters such as fertilizer value (nutrient content), chemical stability or hygienic requirements (absence of pathogens). The introduction of uniform standards is aimed at protecting human, animal, and environmental health while allowing efficient recycling of nutrients.

Heavy metals are one of the main concerns associated with recycling waste into fields. In biomass ash, these metals (e.g., cadmium, lead, mercury, arsenic, chromium, nickel) can occur in concentrations higher than in typical soil, because the combustion process concentrates inorganic elements in the ash residue. To prevent soil degradation and the risk of metal accumulation in the food chain, lawmakers have set maximum permissible concentrations of heavy metals in fertilizers. In the European Union, Regulation 2019/1009 introduced, for the first time, EU-wide limits for contaminants in fertilizer products. Under the 60 mg Cd kg^{-1} P_2O_5 ceiling, a fertilizer containing 45% P_2O_5 may contain up to $\approx$ 27 mg Cd kg^{-1} finished product [10]. This conversion equals only a few mg Cd per kg of finished product, and the limit is meant to curb cadmium inputs from fertilizers because phosphate rock, the raw material for phosphatic fertilizers, frequently carries Cd impurities. Organic and organic-mineral fertilizers (which will include ash and biochar products) also have limits for heavy metals, similar to those previously used in member states. For example, for composts or soil improvers, typical limits in dry mass are on the order of: cadmium ~ 1–3 mg/kg, lead ~ 100–150 mg/kg, mercury ~ 1–2 mg/kg, arsenic ~ 10–13 mg/kg, chromium ~ 90–100 mg/kg, nickel ~ 50 mg/kg, zinc ~ 300–400 mg/kg—the exact values depend on the product category and country. The new EU law harmonizes these values across the Union, making trade and enforcement easier. Importantly, the regulation allows for further tightening of limits in the future on the basis of risk assessments—for example, it is planned to reconsider lowering the permissible level of cadmium in phosphate mineral fertilizers, if available technologies allow it [10].

Biochar as a material added to soil brings other potential quality risks. In addition to heavy metals (the levels of which depend on the raw material—e.g., biochar made from wood tends to have low concentrations of metals, while biochar made from sewage sludge or manure needs to be controlled for possible Cd, Zn, Cu content), a key issue is organic pollutants generated during the pyrolysis process [16]. The most important group are polycyclic aromatic hydrocarbons (PAHs), many of which have carcinogenic or mutagenic properties. The process of producing biochar at low temperatures or with insufficient oxygen can result in tar deposition and PAH formation in the pores of the biochar [17]. EU rules and the EBC/IBI voluntary schemes each cap PAH16 at ≤ 6 mg kg^{-1} for agricultural biochar. As mentioned, the EU has adopted an aggregate limit of PAH16 = 6 mg/kg in biochar used as fertilizer [10]. For comparison, The EBC (European Biochar Certificate) caps the sum of the 16 EPA-PAHs at ≤ 6 mg kg^{-1} for every agricultural grade; the stricter ≤ 4 mg kg^{-1} ceiling applies only to biochar certified for animal feed. No 12 mg kg^{-1} tier exists in the current or previous EBC rules [18]. Among PAHs, benzo(a)pyrene is an indicator of pollution—its contribution to the total of 16 PAHs is often monitored separately. In addition, the ECB and some national guidelines (e.g., in Germany) also include a limit for the sum of 8 EPA-PAHs or EFSA-PAH8 (the most toxic PAHs) at 1 mg/kg [18].

Other potential organic contaminants include dioxins and furans (PCDD/F)—highly toxic compounds that can be formed in thermal processes. Both the EU and the ECB standard require that the content of these compounds in biochar be ultra-low (the mentioned 20 ng TEQ/kg is a very restrictive value, corresponding to trace amounts) eur-lex.europa.eu. Thanks to high pyrolysis temperatures (>500 °C), the realistic dioxin content of biochar from pure biomass usually meets these criteria, but this requirement protects against the use of biochar contaminated with, for example, plastic residues or chlorinated compounds (which promote dioxin formation) [19].

In addition to pollution control, quality standards also address the utility value of these materials. In the case of biomass ash, the content of nutrients such as P, K, Ca, and Mg is important. Some jurisdictions require that ash used as a fertilizer contain minimum concentrations of calcium and potassium, for example, to guarantee a liming or fertilizing effect [20]. Sweden, for example, recommends that ash returned to the forest has a certain minimum Ca + Mg content (for effective soil deacidification) and keeps the ratio of basic cations to heavy metals within safe limits. Biochar, in turn, is evaluated in terms of its organic carbon content (often > 50% C_org), the stability of this carbon (indirectly through the H/C_org < 0.7 ratio), and its effect on soil properties. Standards also require declarations of the content of basic minerals (N, P, K) in biochar, although these are usually low—biochar acts as a carbon carrier and structuring agent rather than as a classic nitrogen or phosphorus fertilizer. Nevertheless, for the sake of transparency and legal compliance, the manufacturer must state the nutrient content if the biochar is marketed as a soil improver [18, 19].

Sanitary (microbiological) safety issues are also important, although in the case of ash and biochar the risk is low due to the high temperatures of the processes. Biomass combustion effectively eliminates pathogens, and ash is a mineral material that is sterile in this regard. Similarly, pyrolysis inactivates pathogens, paving the way for the safe use of biochar even from problematic materials (e.g., sewage sludge or manure), as long as the final product meets chemical requirements [21]. The EU regulations still refer to animal by-product regulations—for example, biochar from manure or bones must additionally meet the requirements of Regulation (EC) No. 1069/2009 (*Animal By-Product (ABP) Regulation*), unless the Commission formally recognizes the pyrolysis process as sanitary (which is under development) [5]. It is also required that raw materials for biochar production not be contaminated with other waste streams (e.g., admixture of plastics, metals) [10].

Existing quality and safety standards provide the foundation for confidence in ash and biochar fertilizers. Multifaceted restrictions—from heavy metals to PAHs and dioxins—ensure that recycling these materials into the soil will not result in risks to human health or the environment. Studies confirm that properly managed ash and biochar can easily meet the adopted limits. For example, analysis of clean biomass ash showed that heavy metal concentrations in all samples tested were below applicable European standards [22]. Nevertheless, maintaining rigorous monitoring is necessary, especially when the raw material is municipal or industrial waste. Standardization of testing methods (e.g., standardizing the extraction of metals from ash) further helps with reliable quality control. The next section of the chapter discusses how these regulations and standards translate into future agricultural policies (Table 7.1).

7.5 Future Prospects for Agricultural Policy

The mainstreaming of biomass ash and biochar is part of broader trends in the transformation of the agri-food sector. The prospects for regulatory and policy development in this area look promising, with key motivations being the move toward a circular economy, the need to ensure food security while reducing dependence on primary raw materials (phosphate, natural gas for nitrogen fertilizer production), and combating climate change [25].

Closing the nutrient cycle: Future agricultural policy will increasingly emphasize the recovery and reuse of nutrients from biomass. The European Commission is already pointing to the need for more efficient management of phosphorus and nitrogen in documents such as the European Green Deal and the Roadmap for a Closed-Circuit Economy [26]. In practice, this means supporting research and investment in technologies for converting organic waste (straw, manure, sewage sludge, wood waste) into safe fertilizers—including ash drying and pelletizing, biochar production, composting, fermentation or struvite recovery. Agricultural policies—such as the Common Agricultural Policy (CAP)—can offer financial incentives to farmers using secondary fertilizers. Already, some member countries are envisioning measures to promote bio-based fertilizers in their Strategic Plans for the CAP. Farm to Fork (Farm to Table strategy) also envisages reducing nutrient losses and developing alternative fertilizers [27]. Further updates to directives (e.g., Sludge Directive 86/278/EEC on sewage sludge) or new legislative initiatives can be expected to set specific phosphorus recycling targets at the EU level, which will indirectly stimulate the use of ash and biochar rich in this element.

Innovation and scientific research: Looking ahead, regulations will be based on the latest scientific knowledge. Research projects are currently underway to assess

Table 7.1 Examples of regulations and standards for biomass ash and biochar in agriculture

Regulation/standard	Scope and legal status	Key provisions
Regulation (EU) 2019/1009 on fertilizing products [2, 10]	EU law (applicable since 16 July 2022)	Allows placing ash (CMC 13) and pyrolysis materials/biochar (CMC 14) on the EU market as CE-marked fertilising products when criteria are met. Sets contaminant limits (Annex I) and defines raw-material and process requirements (Annex II). Products that fully meet the Regulation's requirements can be placed on the market and may achieve end-of-waste status under the harmonised framework
Waste framework directive 2008/98/EC [3]	EU law (waste)	Defines waste, by-products, and criteria for end-of-waste status. Ash or biochar derived from waste remains classified as waste until recovered according to the law or meets the conditions of the fertilizer regulation. National permits (e.g., for R10 process—land spreading of waste) may allow local agricultural use of ash
European Biochar Certificate (EBC) [18]	Voluntary standard (Europe)	Defines biochar and quality classes (agro, feed, etc.). Limits heavy metals (e.g., Cd < 1.5 mg/kg, Pb < 150 mg/kg for agro class) and sum of 16 PAHs (<6 mg/kg). Requires $H/C_{org} < 0.7$, defines test methods and composition declaration requirements. Complements legal regulations—certified biochar more easily meets legal requirements
IBI biochar standards (International Biochar Initiative) [23]	Voluntary standard (global)	Similar scope to EBC: standard definition of biochar, lab testing protocols, quality criteria (metals, PAHs, pH, carbon content, physical parameters). Serves as a guide for non-EU producers and facilitates product comparability in the global market
National regulations—example: Poland [24]	National law (waste/fertilizers)	In Poland, there is no separate "ash fertilizer" category—biomass ash is considered waste. The Ministry of Environment regulation (2015) governs R10 recovery (land application), requiring, among others, compliance with heavy metal limits in soil after application. Since 16 July 2022, products complying with EU 2019/1009 (including CMC 13/14) can be placed on the market with CE marking; otherwise, national waste rules continue to apply

the long-term effects of adding biochar and ash to soil, including their impact on fertility, nitrous oxide emissions, carbon storage, and soil microbiome health. The results of this research may lead to further adjustments to legislation: for example, if it turns out that certain contaminants do not accumulate in crops, the limits may be relaxed or, conversely, tightened if undesirable side effects are detected. There are already indications of a need to improve the risk assessment framework for new biofertilizers. Silva et al. suggested that revisions to the risk assessment approach will be necessary to cover the specific characteristics of materials such as biomass ash [1]. The same applies to the standardization of analytical methods—future standards are likely to refer to international standards (e.g., ISO standards) currently being developed for biochar and biocoal. In March 2025, ISO established an expanded Technical Committee ISO/TC 238 on biochar and pyrolytic biofuels to develop harmonized terminology, classifications, and test methods for biocharubiochareurope.eubiochareurope.eu. This will facilitate international trade and the recognition of test results between countries, and may also be adapted to legal regulations (so-called harmonized standards). In the future, we may see references to ISO standards in the context of testing, e.g., for PAH content or phosphorus availability in such products [28].

Climate policy and carbon farming: Biochar in particular has an important role to play in climate policy. Since the carbon contained in biochar can remain in the soil for hundreds of years, its application is considered a method of carbon dioxide removal (CDR). The European Union has recognized this potential and is currently preparing a Carbon Dioxide Removal Certification Framework (CRCF). In November 2022, the European Commission presented a proposal for a regulation on carbon dioxide capture certification, which provides for three main removal pathways: carbon farming (e.g., increasing soil organic matter content), permanent storage in products, and geological storage. Biochar fits into the first two categories—as a carbon farming practice (a soil additive that increases stable carbon content) and as a potential component of long-term carbon sequestration products. In 2024, special EU workshops and consultations are being held on the methodology for certifying carbon sequestration using biochar [29]. It is expected that official guidelines on how to document and verify carbon credits for the use of biochar will be developed in the near future. For farmers and businesses, this could open up a new source of income (selling certificates for CO_2 absorption) while strengthening the case for public support. Tax breaks (e.g., the 45Q credit) have already been adopted in the US, and there is ongoing debate about legislation promoting biochar in agriculture, while in Europe, pilot programs (e.g., under Horizon Europe) are funding carbon farming activities using biochar [30]. The trend is clear: carbon sequestration will become an integral part of agricultural policy, and biochar is one of the most promising tools for this purpose. This is conditional on maintaining high environmental standards—hence certification will be strictly linked to prior compliance with quality standards (there will be no "carbon credit" for biochar that is not approved as a safe soil improver) [31].

Sustainable agriculture and soil health: In the future, policy will place greater emphasis on a holistic approach to soil fertility. The introduction of ash and biochar may be part of regenerative agriculture practices, improving soil structure, increasing water capacity, and biological activity. It is possible that agri-environmental programs will include subsidies for farms that use such practices. At the global level, the UN Sustainable Development Goals (Agenda 2030) for soil protection and food security support initiatives that reduce soil degradation and resource waste. Recycling ash (instead of landfilling it) and producing biochar from waste biomass (instead of burning it in fields or allowing it to rot and emit methane) are examples of actions that fit these goals. We can therefore expect an increase in international cooperation and the exchange of good practices. Organizations such as the FAO are already publishing reports and guides on the safe use of residual products (e.g., biosolids, i.e., sludge, and ash) in agriculture. Over time, these recommendations may take the form of soft law or standards that countries will implement [1, 11, 32].

Market development and education: The last but important aspect of the future is shaping the market for ash and biochar fertilizers. Policy can support demand for these products, for example through public procurement requirements (preference for recycled fertilizers in public procurement for green areas, etc.) or through educational campaigns to raise awareness among farmers of the benefits. There will likely be a need for further standardization, e.g., creating a system for labeling the nutrient content and neutralization value on such fertilizers so that users know exactly what they are buying. Regulation 2019/1009 already requires fertilizers containing ash or biochar to be labeled with information on, nutrient content, pH, electrical conductivity, bulk density, etc., if relevant. Standardizing these labels will make it easier to compete with conventional fertilizers. The future will likely also see the emergence of specialized companies involved in collecting waste biomass and producing fertilizers from it-policy can support this sector, for example, with innovation funds [33, 34].

In summary, the outlook is optimistic: agriculture of the future is likely to incorporate biomass ash and biochar into mainstream agricultural practices as part of a broader strategy for sustainable resource management. The key pillars are: refined regulations (which will evolve with scientific knowledge), effective quality standards that guarantee safety, market mechanisms (including payments for ecosystem services such as carbon sequestration), and continuous awareness-raising among stakeholders. This chapter shows that a big step forward has already been taken-from a policy and legal perspective, the foundations have been laid to close the cycle of matter from the field, through biomass, to ash, and back to the field in the form of fertilizer. Challenges remain, but the direction of change is consistent with the idea of environmentally friendly agriculture based on the circulation of nutrients and the responsible use of new materials for the benefit of soil and climate. All this allows us to look to the future of agricultural policy with cautious optimism in the context of biomass ash and biochar.

References

1. F.C. Silva, N.C. Cruz, L.A.C. Tarelho, S.M. Rodrigues, Use of biomass ash-based materials as soil fertilisers: critical review of the existing regulatory framework. J. Clean. Prod. **214**, 112–124 (2019). https://doi.org/10.1016/J.JCLEPRO.2018.12.268
2. E. Jarosz-Krzemińska, J. Poluszyńska, Repurposing fly ash derived from biomass combustion in fluidized bed boilers in large energy power plants as a mineral soil amendment. Energies **13**(18), 4805 (2020). https://doi.org/10.3390/EN13184805
3. *Directive 2008/98/EC of the European Parliament and of the Council of 19 November 2008*. Accessed 11 Sept 2022. [Online]. Available: https://eur--lex-europa-eu.translate.goog/legal-content/EN/TXT/?uri=celex:32008L0098&_x_tr_sl=en&_x_tr_tl=pl&_x_tr_hl=pl&_x_tr_pto=sc
4. A. Maresca, J. Hyks, T.F. Astrup, Recirculation of biomass ashes onto forest soils: ash composition, mineralogy and leaching properties. Waste Manage. **70**, 127–138 (2017). https://doi.org/10.1016/j.wasman.2017.09.008
5. M. Álvarez Salas et al., Current challenges on the widespread adoption of new bio-based fertilizers: insights to move forward toward more circular food systems. Front. Sustain. Food Syst. **8**, 1386680 (2024). https://doi.org/10.3389/FSUFS.2024.1386680/XML/NLM
6. A. Uliasz-Bocheńczyk, E. Mokrzycki, The elemental composition of biomass ashes as a preliminary assessment of the recovery potential. Gospod. Surowcami Miner./Miner. Resour. Manag. **34**(4), 115–132 (2018). https://doi.org/10.24425/122593
7. *Labeling Guidelines for Biochar Products | US Biochar Initiative*. Accessed 29 June 2025. [Online]. Available: https://biochar-us.org/labeling-guidelines-biochar-products
8. C.O. Rusănescu, M. Rusănescu, Application of fly ash obtained from the incineration of municipal solid waste in agriculture. Appl. Sci. **13**(5), 3246 (2023). https://doi.org/10.3390/APP13053246
9. I. Nogués et al., Biochar soil amendment as carbon farming practice in a Mediterranean environment. Geoderma Reg. **33**, e00634 (2023). https://doi.org/10.1016/J.GEODRS.2023.E00634
10. *Regulation (EU) 2019/1009 of the European Parliament and of the Council of 5 June 2019 Laying Down Rules on the Making Available on the Market of EU Fertilising Products and Amending Regulations (EC) No 1069/2009 and (EC) No 1107/2009 and Repealing Regulation (EC) No 2003/2003*. Accessed 26 Oct 2022. [Online]. Available: https://eur-lex.europa.eu/legal-content/EN/TXT/PDF/?uri=CELEX:32019R1009&from=EN
11. S.R. Kumar, Biochar-boon to soil health and crop production. Afr. J. Agric. Res. **8**(38), 4726–4739 (2013). https://doi.org/10.5897/AJAR2013.7239
12. F. Danso, W.A. Agyare, A. Bart-Plange, Benefits and costs of cultivating rice using biochar-inorganic fertilizer combinations. J. Agric. Food Res. **11**, 100491 (2023). https://doi.org/10.1016/J.JAFR.2022.100491
13. S.K. Das, G.K. Ghosh, R. Avasthe, Application of biochar in agriculture and environment, and its safety issues. Biomass Convers. Biorefin. **13**(2), 1359–1369 (2023). https://doi.org/10.1007/S13399-020-01013-4/FIGURES/1
14. A.K. Sakhiya, A. Anand, P. Kaushal, Production, activation, and applications of biochar in recent times. Biochar **2**(3) (2020). https://doi.org/10.1007/s42773-020-00047-1
15. D. Cao et al., Production equipment and production process of biochar-based compound fertilizer. Patent no. CN111392711B, Nov 2020
16. J. Chew et al., Biochar-based fertilizer: supercharging root membrane potential and biomass yield of rice. Sci. Total Environ. **713** (2020). https://doi.org/10.1016/J.SCITOTENV.2019.136431
17. C. Wang et al., Biochar-based slow-release of fertilizers for sustainable agriculture: a mini review. Environ. Sci. Ecotechnol. **10**, 100167 (2022). https://doi.org/10.1016/J.ESE.2022.100167
18. European Biochar Certificate, *Guidelines for Biochar Production for a Sustainable Production of Biochar*

19. Ł. Sobol, A. Dyjakon, K. Soukup, Dioxins and furans in biochars, hydrochars and torreficates produced by thermochemical conversion of biomass: a review. Environ. Chem. Lett. **21**(4), 2225–2249 (2023). https://doi.org/10.1007/S10311-023-01600-7
20. A. El-Naggar et al., Biochar composition-dependent impacts on soil nutrient release, carbon mineralization, and potential environmental risk: a review. J. Environ. Manage. **241**, 458–467 (2019). https://doi.org/10.1016/J.JENVMAN.2019.02.044
21. W. Suliman, J.B. Harsh, A.M. Fortuna, M. Garcia-Pérez, N.I. Abu-Lail, Quantitative effects of biochar oxidation and pyrolysis temperature on the transport of pathogenic and nonpathogenic *Escherichia coli* in biochar-amended sand columns. Environ. Sci. Technol. **51**(9), 5071–5081 (2017). https://doi.org/10.1021/ACS.EST.6B04535
22. M. Pazalja et al., Heavy metals content in ashes of wood pellets and the health risk assessment related to their presence in the environment. Sci. Rep. **11**(1), 1–9 (2021). https://doi.org/10.1038/S41598-021-97305-4
23. International Biochar Initiative. Accessed 29 June 2025. [Online]. Available: https://biochar-international.org/
24. Polska Unia Ubocznych Produktów Spalania. Accessed 29 June 2025. [Online]. Available: http://unia-ups.pl/
25. P. Sashidhar et al., Biochar for delivery of agri-inputs: current status and future perspectives. Sci. Total Environ. **703**, 134892 (2020). https://doi.org/10.1016/J.SCITOTENV.2019.134892
26. B.R. Kiran, M.N.V. Prasad, S.V. Mohan, Farm to fork: sustainable agrifood systems, in *Sustainable and Circular Management of Resources and Waste Towards a Green Deal* (2023), pp. 25–38. https://doi.org/10.1016/B978-0-323-95278-1.00012-7
27. I.A. Elshaer, A.M.S. Azazz, S.S. Hassan, S. Fayyad, Farm-to-fork and sustainable agriculture practices: perceived economic benefit as a moderator and environmental sustainability as a mediator. Sustainability **15**(14), 11462 (2023). https://doi.org/10.3390/SU151411462
28. W. Yang et al., PAHs sorption to biochar colloids changes their mobility over time. J. Hydrol. (Amst.) **603**, 126839 (2021). https://doi.org/10.1016/J.JHYDROL.2021.126839
29. *Carbon Removals and Carbon Farming—European Commission*. Accessed 29 June 2025. [Online]. Available: https://climate.ec.europa.eu/eu-action/carbon-removals-and-carbon-farming_en
30. *The Regulatory Landscape for Biochar: Highlights from 2024 and What to Expect in 2025—Biochar Today*. Accessed 29 June 2025. [Online]. Available: https://biochartoday.com/blog/the-regulatory-landscape-for-biochar-highlights-from-2024-and-what-to-expect-in-2025/
31. F.U. Haider et al., Co-application of biochar and microorganisms improves soybean performance and remediate cadmium-contaminated soil. Ecotoxicol. Environ. Saf. **214**, 112112 (2021). https://doi.org/10.1016/J.ECOENV.2021.112112
32. M. He et al., A critical review on performance indicators for evaluating soil biota and soil health of biochar-amended soils. J. Hazard. Mater. **414**, 125378 (2021). https://doi.org/10.1016/J.JHAZMAT.2021.125378
33. A.E. Latawiec et al., Analysis of the economic potential trough biochar use for soybean production in Poland. Agronomy **11**(11), 2108 (2021). https://doi.org/10.3390/AGRONOMY11112108/S1
34. M. Owsianiak, H. Lindhjem, G. Cornelissen, S.E. Hale, E. Sørmo, M. Sparrevik, Environmental and economic impacts of biochar production and agricultural use in six developing and middle-income countries. Sci. Total Environ. **755**, 142455 (2021). https://doi.org/10.1016/J.SCITOTENV.2020.142455

Chapter 8
Sustainable Development and Environmental Protection

8.1 Introduction

Ash from biomass combustion and biochar produced in the biomass pyrolysis process are becoming important elements of the circular economy and sustainable development. Biochar is recognized as a carbon dioxide removal (CDR) technology that promotes food security, sustainable land management, and the circular economy [1]. Ash, on the other hand, is an unavoidable by-product of thermal biomass processing and is often still treated as waste stored in landfills—currently, only a small fraction of biomass ash is recycled [2]. Meanwhile, both biochar and ash can be reused in ways that bring significant environmental and climate benefits. This chapter discusses how the use of ash and biochar fits into the goals of sustainable development and environmental protection. It presents their role in reducing greenhouse gas emissions and carbon sequestration, their importance in waste management and biomass recycling, as well as the long-term environmental benefits of their use. Particular emphasis is placed on current global trends and research findings in this area.

8.2 Greenhouse Gas Emissions and Carbon Sequestration

Biochar production and biomass ash utilization are strategies that can contribute to climate change mitigation by reducing greenhouse gas emissions and permanently sequestering carbon. Biochar is produced by pyrolyzing biomass (e.g., agricultural waste and wood residues) under conditions of limited oxygen access. In this process, a significant portion of the carbon contained in the biomass remains in the solid product—biochar—instead of escaping as CO_2 [3]. The resulting biochar has a very high content of stable elemental carbon, allowing it to serve as a long-term carbon store in the ecosystem. Scientific estimates indicate that the global emission reduction potential of biochar systems reaches gigatons of cumulative CO_2 per year—according

K. Chojnacka et al., *Sustainable Fertilizers: Utilizing Biomass Ashes and Biochars*, SpringerBriefs in Applied Sciences and Technology,
https://doi.org/10.1007/978-3-032-03765-7_8

to a review of 19 studies, from about 0.03 to as much as 11 Pg CO_2 equivalent per year [1]. Such a wide range is due to different assumptions about biomass availability and technology, but even conservative, independent analyses suggest a possible sustainable emission reduction potential in the range of 2.6–10.3 Pg CO_2/year. To illustrate the scale, the upper limit of these estimates corresponds to more than 20% of global annual greenhouse gas emissions [1].

8.2.1 Carbon Sequestration in Biochar

The key mechanism behind biochar's beneficial impact on the greenhouse gas balance is the sequestration of atmospheric carbon in a stable form. The carbon contained in biochar is resistant to biodegradation—it can remain in the soil for hundreds or even thousands of years, which means that CO_2 is effectively removed from the atmospheric cycle on a centuries-long scale [4]. By comparison, the carbon contained in raw biomass (e.g., plant residues) would be released through decomposition or combustion within months or a few years. The durability of biochar is confirmed by both laboratory tests and field observations—for example, the discovery of ancient terra preta soils in the Amazon, enriched with carbonaceous matter hundreds of years ago, testifies to the long-term stability of carbon in this form. Biochar therefore acts as a carbon dioxide sink, trapping carbon from atmospheric CO_2 (previously assimilated by plants) in the soil for a very long time [3].

8.2.2 Reduction of N_2O and CH_4 Emissions

In addition to direct carbon sequestration, biochar also reduces emissions of other greenhouse gases from ecosystems, especially nitrous oxide (N_2O) and methane (CH_4) [5]. The addition of biochar to soil changes its physicochemical properties—it increases pH, improves aeration (air access), and soil structure—which affects the microbial processes responsible for N_2O and CH_4 emissions [5]. Numerous studies have shown that biochar reduces nitrous oxide emissions from agricultural land by stimulating complete denitrification (reduction of N_2O to N_2) and sorption of available forms of nitrogen in the soil. Meta-analyses indicate that, on average, the addition of biochar reduces N_2O emissions by approximately 30–50% [6]. For example, a global quantitative analysis showed an average decrease in N_2O emissions of 38% after applying biochar to soilsfrontiersin.org. Individual experiments reported even greater reductions—in the range of 80–90%—at high doses of biochar [6]. The reduction in N_2O emissions results, among other things, from an increase in soil pH (greater availability of N_2O-reducing enzymes) and the sorption of ammonium and nitrate ions by porous biochar, which reduces the substrates for denitrification processes [7]. In the case of methane (CH_4), the effect of biochar is most noticeable in wet soil conditions (e.g., rice cultivation). The addition of biochar to flooded rice soils improves their oxidation, which limits methanogenesis. A research field in the Yangtze River basin (China) showed that in a biochar-enriched rice paddy, total CH_4 and N_2O emissions decreased significantly compared to the control, while long-term organic carbon in the soil increased. Overall, biochar reduces GHG emissions per unit of yield and lowers the global warming potential (GWP) [8].

8.2.3 Replacing Fossil Fuels and Avoiding Emissions from Waste

In both biochar systems and biomass combustion for energy purposes, an important factor in reducing overall emissions is the use of biomass as a renewable fuel instead of fossil fuels. Biomass is considered a CO_2-neutral fuel because the CO_2 released during its combustion was previously absorbed from the atmosphere by plants (over years, not millions of years as in the case of fossil fuels) [9]. Replacing coal or natural gas with biomass in energy production therefore leads to a significant net reduction in greenhouse gas emissions. In addition, the pyrolysis process to produce biochar yields synthesis gas (syngas) and liquid (bio-oil) as by-products, which can be used for energy. Burning syngas from pyrolysis to produce energy means that some of the chemical energy of the biomass is used, reducing the demand for fossil fuels [10]. Furthermore, this approach eliminates emissions that would arise from alternative treatments of biomass waste. Leaving plant residues in the field to decompose or burning them uncontrollably would result in the release of the carbon they contain as CO_2 or CH_4. By converting this waste into biochar or energy, we avoid these emissions [11]. As analyses show, biochar systems combine these numerous climate benefits—permanent carbon sequestration, reduction of soil N_2O and CH_4 emissions, replacement of fossil fuels with renewable energy from syngas, and avoidance of emissions from rotting or burning residues [1].

8.3 The Role of Ash in the Emissions Balance

Although biomass ash itself does not contain significant amounts of carbon (most carbon burns to CO_2), its skillful use can also contribute to reducing greenhouse gas emissions indirectly. One way is to use ash in the building materials industry as a substitute for carbon-intensive raw materials [12, 13]. For example, the addition of ash to cement or the production of so-called geopolymers with ash makes it possible to reduce the use of Portland clinker, the production of which is associated with high CO_2 emissions. Replacing 1 ton of cement clinker with ash means avoiding several hundred kilograms of CO_2 emissions that would have been generated by firing carbonates and fossil fuels in a cement plant [14]. In addition, biomass ash is rich in calcium compounds (CaO, $Ca(OH)_2$), which exhibit the ability to bind CO_2

from the air through carbonation reactions. Research indicates that alkaline ash can capture CO_2 from the atmosphere to form stable calcium carbonates and magnesium [2]. The process is being explored as a low-cost form of carbon capture and storage (CCS)—under controlled conditions of accelerated carbonation, it is possible to bind up to tens of kilograms of CO_2 per ton of ash, while stabilizing potentially harmful ash components in the carbonate matrix. Although ash carbonization is mainly relevant on an experimental scale for now, it is an interesting complement to CO_2 efforts to reduce CO_2 [15]. Table 8.1 summarizes the key characteristics of biochar and ash from the perspective of their impact on GHG emissions and carbon sequestration.

8.4 Energy Use of Waste Biomass

One way to reduce the amount of organic waste is to use it to produce energy. Burning wood, agricultural, or municipal waste under controlled conditions allows energy to be recovered in the form of heat and electricity, while reducing the volume of waste sent to landfills [16]. For example, many countries promote the co-combustion of biomass in power plants as a form of simultaneous renewable energy production and organic waste disposal [17]. The ashes produced in these processes still contain valuable minerals, which is why applications for them are increasingly being sought instead of depositing them in landfills. Unfortunately, on a global scale, a large proportion of ash is still landfilled—for example, in Canada, landfilling remains the most common method of ash disposal [18]. However, global trends are moving towards increasing the proportion of ash that is recycled. Legislation in many countries (including the European Union) is gradually tightening restrictions on the disposal of combustion waste, which is forcing the development of technologies for its utilization [19].

Biochar from organic waste. Another approach to the problem of biomass waste is pyrolysis technology, which produces biochar. This process is particularly attractive for waste that is difficult or undesirable to burn directly, such as crop residues, straw, food processing waste, and even sewage sludge. Converting such substrates into biochar reduces their mass and volume, destroys pathogens and organic pollutants (e.g., pharmaceuticals in sewage sludge), and at the same time retains some of the elements in the solid product [20]. Biochar obtained from waste can be used as a soil improver or pollutant sorbent, closing the material cycle. For example, programs are being developed in China for the pyrolysis of crop residues (e.g., rice straw), which avoids their widespread burning in fields, causing smoke and emissions—instead, biochar is produced, improving soil fertility and storing carbon. In this way, the problem of burdensome agricultural waste has been solved and a product that promotes climate protection has been obtained [21].

Table 8.1 Comparison of the environmental properties and functions of biochar and biomass ash

Property	Biochar	Biomass ash
Production method	Pyrolysis of biomass in anaerobic conditions (approx. 300–700 °C)—partial carbonization of organic material	Combustion of biomass in the presence of oxygen (boilers, furnaces, etc.)—complete oxidation of organic matter to CO_2, residual is inorganic ash
Carbon content	Very high (carbonized biomass contains 50–90% C by mass). Most of the carbon remains in the solid product	Negligible (carbon in biomass is burned to CO_2; ash is mainly composed of oxides and mineral salts, < 5% C)
Stability in the environment	High—carbon in biochar is resistant to biological degradation, stable for hundreds of years in soil (long-term CO_2 sequestration)	Low in terms of carbon—lacks stable carbon for sequestration; mineral components of ash may undergo slow changes (dissolution, weathering) in the environment
pH level	Neutral to alkaline (typically pH ~ 7–10, depending on feedstock and pyrolysis temperature). Mitigates soil acidification	Strongly alkaline (often pH 11–13 in the case of wood ash). Strongly reduces soil acidity (acts like lime)
Nutrient content	Contains some macro- and micronutrients (K, P, Ca, Mg, trace elements), but often in less available forms. Acts like a sponge absorbing nutrients, reducing their loss	Rich in mineral ash: high content of calcium, potassium, phosphorus, magnesium and other elements. Nutrients are more soluble—can quickly supply the soil with available forms (fertilizing effect)
Impact on soil	Improves soil structure (due to porosity and stability of biochar particles), increases water holding capacity and air availability. Promotes soil microbial activity	Raises soil pH (neutralizes acidity), provides plants with available mineral nutrients (acts like a lime-potassium fertilizer). Excessive amounts may cause soil salinity or alkalization
Main climate benefits	Permanent carbon storage (from atmospheric CO_2) in soil; reduction of N_2O and CH_4 emissions from cultivated soils; indirectly—avoids emissions from biomass decay and replaces fossil fuels with syngas energy	Indirect reduction of CO_2 emissions through: (a) replacing cement clinker and lime (reducing industrial process emissions), (b) recycling nutrients (less need for energy-intensive synthetic fertilizers), (c) potential CO_2 capture through carbonation (binding carbonates in ash)

(continued)

Table 8.1 (continued)

Property	Biochar	Biomass ash
Other environmental benefits	Improves soil fertility and crop yield (water retention, nutrient storage); reduces nitrate leaching into water bodies; reclamation of degraded soils (binding pollutants like heavy metals and improving organic matter)	Neutralization of process residues (ash is reused instead of landfilled); saving natural resources (sand, cement, phosphate rock); formation of stable mineral matrices that immobilize potential pollutants (heavy metals) in construction materials
Potential challenges	Requires quality control (possible presence of PAHs at too low pyrolysis temperatures; potential contaminants from feedstock). Determining optimal doses—excess biochar may disrupt soil balance (e.g., block availability of certain nutrients)	Composition variability depending on biomass type and combustion process (hinders application standardization). Risk of elevated heavy metals (especially in ash from waste, bark, straw)—requires monitoring to ensure safe application. High alkalinity and salinity may limit soil application rates

8.5 Nutrient Recycling

Both ash and biochar are forms of condensed minerals derived from biomass. They are concentrated almost entirely in the ash—plant nutrients (calcium, potassium, phosphorus, magnesium, silicon, and others) remain in the residue after combustion. Biochar retains some of these elements (the rest ends up in the ash fraction in bio-oil and syngas), but because it is applied to the soil in large doses, it also contributes significant amounts of macro- and microelements to it [22]. The reintroduction of these substances into fields or forests closes the biogenic element cycle. This is particularly important in the context of phosphorus, an element critical for food production, whose geological resources (phosphorites) are limited and subject to depletion. Phosphorus-rich ashes can serve as an alternative, sustainable source of phosphorus fertilizers, reducing agriculture's dependence on fossil raw materials [23]. The introduction of wood ash into farmland or forests also replenishes potassium and calcium, acting similarly to traditional liming and potassium fertilization [24]. Studies have shown that the use of ash reduces the consumption of artificial fertilizers, contributing to savings in raw materials and energy needed for their production. In Scandinavian countries, wood ash has been returned to forests for years as part of so-called ash recycling—this is to compensate for the removal of nutrients along with harvested wood and to prevent the acidification of forest soils [24].

8.6 Applications in Agriculture and Reclamation

Biomass ash can be a valuable calcium-potassium fertilizer for agricultural land if it meets purity standards. The introduction of moderate doses (e.g., 1–3 t/ha) of wood ash to soils poor in potassium and phosphorus usually results in an increase in the soil's content of these elements and an increase in pH in acidic soils [25]. This translates into improved crop yields, especially in conditions of potassium deficiency or on acidified soils. Ashes are also used for the reclamation of degraded areas, e.g., after mining or at landfills. The calcium oxides and hydroxides they contain effectively raise the pH of the substrate, which helps neutralize acidified post-mining soils [26]. At the same time, heavy metals are stabilized—in an alkaline environment, many metals cease to be mobile. Biochar is also used in reclamation—thanks to its high porosity and specific surface area, it can bind organic toxins and certain heavy metals, limiting their bioavailability in contaminated soils. In this way, both materials, derived from waste, are used to neutralize other waste or pollutants through immobilization and improvement of soil quality [27].

8.7 Challenges and Safety of Use

It should be emphasized that the recycling of ash and biochar must be carried out with caution and in compliance with environmental standards. Ash may contain potentially toxic elements (such as cadmium, lead, zinc), especially if it comes from the combustion of contaminated biomass or mixed waste [28]. Composition studies show that heavy metals are usually found in the highest concentrations in fly ash, while bottom ash contains more macronutrients and fewer volatile metals. Nevertheless, any type of ash intended for use as fertilizer should undergo chemical analysis [29]. It is necessary to monitor the content of heavy metals and other contaminants so that the application of ash does not cause the accumulation of toxic substances in the soil. Many countries have regulations specifying the permissible doses and quality of ash used in agriculture (e.g., limits for Cd, Pb, Hg, As), which protects against negative impacts on the environment and health. In the case of biochar, however, it is important to produce it from clean, toxin-free raw materials and under appropriate conditions—too low a pyrolysis temperature can result in the presence of tarry compounds such as PAHs (polycyclic aromatic hydrocarbons). Quality standards, such as the European Biochar Certification (EBC), specify the permissible levels of contaminants in biochar intended for use in soil. Compliance with these standards guarantees that the use of biochar and ash will be safe and beneficial [30].

8.8 Industrial and Innovative Applications

In addition to agriculture and land reclamation, biomass ash is increasingly being used in the construction materials industry. Already, several percent of Portland cement globally are being replaced by fly ash (mainly from coal, but increasingly also from biomass) and slag, which significantly reduces the carbon footprint of concrete. Wood ash is being tested as an additive to concrete and asphalt mixtures—for example, adding 5–10% wood ash to asphalt improved its mechanical properties and durability [31]. Another direction is the production of geopolymers and ceramic materials with the use of ash, which can replace traditional fired materials (bricks, lightweight aggregates). Importantly, such applications enable the utilization of large amounts of ash, reducing the pressure on its storage. Ash becomes a raw material, not waste—this is a model example of a circular economy [32] (Table 8.2).

The widespread use of biochar and biomass ash has a number of long-term environmental benefits beyond the immediate effects described in the previous sections. These are related to both climate protection and a lasting increase in the quality of the soil environment and resource efficiency.

Table 8.2 Examples of biomass ash applications and their environmental benefits

Biomass ash application	Environmental and sustainable development benefits
Soil fertilization in agriculture and forestry	Contains macronutrients (P, K, Ca, Mg)—can partially replace mineral fertilizers derived from fossil resources (reducing phosphate mining); improves soil fertility and pH (liming effect); in forests, compensates for nutrient removal with biomass, preventing forest habitat degradation
Additive to construction materials (cement, concrete, geopolymers)	Pozzolanic properties of ash allow for partial replacement of cement clinker, reducing CO_2 emissions in the cement industry (less limestone and fuel consumption during clinker production). Using ash as aggregate or additive in concrete and geopolymers limits new raw material extraction (sand, gravel) and reduces landfill waste
Innovative uses (CO_2 capture, catalysts)	High alkaline oxide content (e.g., CaO) enables CO_2 capture through carbonation—ash absorbs CO_2, forming stable carbonates, contributing to carbon sequestration. Ash is also used as a catalyst (e.g., in biodiesel production from waste oils, replacing conventional bases) and as a sorbent (for removing contaminants from water and gases). These innovations expand ash utilization potential and reduce the need for conventional chemicals, supporting sustainable chemistry goals

8.9 Sustained Improvement in Soil Fertility and Function

One of the most important long-term effects of adding biochar to land is the improvement of soil quality, which translates into a number of positive environmental phenomena. Biochar, due to its porosity and high specific surface area, acts as a sponge and structural framework in the soil: it improves the tuberous structure, increases the soil's ability to retain water and the availability of air in the soil profile. In the long term, this means less susceptibility of the soil to erosion (soil with better structure is not as easily washed away or blown away) and greater resistance to drought due to higher water retention [8]. Field studies have shown that a single application of biochar can provide a multi-year increase in water holding capacity and soil organic matter content, which is crucial in an era of more frequent droughts [8]. In addition, biochar promotes the growth of beneficial soil microflora and earthworms, which further improve soil structure and fertility [33]. Taken together, this leads to a sustained increase in ecosystem productivity—in field experiments, increased yields are often observed to persist for many years after a single application of biochar [8]. For example, a 10-year experiment in a vineyard in Italy reported that soil organic carbon content was still significantly higher a decade after biochar was added, and soil fertility was improved (higher nutrient abundance, enzymatic activity) compared to the control. Long-term soil fertility is not only agriculturally important, but also ecologically important—healthy soils store water and carbon better, support microbial biodiversity, and require less fertilizer and pesticide inputs, reducing chemical pressure on the environment [34].

8.10 Carbon Sequestration and Climate Change Mitigation

Permanently binding carbon as biochar means that it will be out of the rapid carbon cycle for decades or centuries. Every ton of carbon stored as biochar in the soil equals a ton less carbon dioxide in the atmosphere (ignoring small losses in the production process, which can be offset by using syngas energy). On a global scale, if biochar technology is applied more widely, we can talk about a significant element in the portfolio of climate neutrality measures [35]. The IPCC organization already includes biochar as one of the options for negative emissions (negative emissions technologies) in climate stabilization scenarios. In doing so, biochar has a relatively high technological readiness (TRL 8–9) compared to other CO_2 biochar-international.org capture methods. What's more, the scaling of biochar production is already underway—in 2023, the biochar industry reached a global production of about 350,000 tons of biochar per year, with a market value of $600 million [36]. The progression is exponential (average annual production growth of 91% between 2021 and 2023), and analyses indicate a viable path to reach absorption of about 1 gigaton of CO_2 per year by 2040 via biochar-biochar-international.org. Already, biochar dominates the volume of CDR carbon credits sold—it accounted for more than 90% of the volume of verified carbon removals sold in the carbon market in 2023.biochar-international.org. This demonstrates growing global confidence and interest in this technology as an effective and sustainable way to neutralize emissions [36].

In the case of biomass ash, its contribution to climate change mitigation is indirect but systemically important. Replacing raw materials and products with a large carbon footprint (such as cement, fertilizers) with materials derived from ash translates into avoided emissions in other sectors of the economy [37]. In addition, extending the life cycle of elements (such as P, K, Ca) through ash recycling means less need to extract and process new resources, which always involves emissions and environmental interference. The long-term benefit, therefore, is a reduced burden on ecosystems from mining and waste activities. It is worth adding that the binding of some of the carbon in the mineral form of carbonates (through the natural carbonation of ashes dispersed in soil or used in building materials) also provides a sustainable sink for CO_2—admittedly on a smaller scale than biochar, but with the mass use of ashes the effect is not negligible [15].

8.11 Supporting the Circular Economy

Biochar and ash fit into the idea of a circular economy, where waste from one process becomes a raw material for another. Thanks to them, it is possible to close the material cycle in the bioenergy and agricultural sectors. Instead of losing valuable components along with waste (gas emissions or storage of solid residues), they return to the production ecosystem. This model reduces the negative impact of agriculture and energy on the environment, making them more sustainable. As stated in recent publications, the use of biochar and ash can play an integral role in implementing sustainable development strategies and building a circular bioeconomy. For example, the implementation of ash recycling in biomass energy makes the entire energy production process a cycle—from biomass, through energy and ash, back to the soil as the place of biomass production. The same applies to biochar: incorporating it into the soil closes the carbon cycle ($CO_2 \rightarrow$ biomass $\rightarrow$ biochar $\rightarrow$ soil) and replenishes nutrient cycles [1, 38, 39].

8.12 Support for Biodiversity and Ecosystem Services

In the long term, measures such as improving soils with biochar or fertilizing with ash can promote local biodiversity. Soil that is richer in organic carbon and minerals creates a better habitat for soil fauna (e.g., earthworms, bacteria, symbiotic fungi) [40]. Healthy soil also means healthier plants—more resistant to disease and abiotic stress, which can reduce the need for chemical plant protection products. In forests, the return of ash can help the revival of plant species that require higher pH values, counteracting excessive acidification of habitats caused by acid deposition and biomass removal. Biochar increases water retention in terrestrial ecosystems, mitigating the effects of droughts and floods (by increasing soil absorbency). This is an example of ecosystem services provided by sustainable practices—we protect the climate, soil, and water resources at the same time [33, 41].

The use of ash and biochar is a modern approach to resource management, combining the goals of climate protection, effective waste management, and environmental improvement. These benefits are not short-term—on the contrary, they are cumulative and long-term. Every batch of carbon stored in the soil as biochar is a gain for the climate for hundreds of years. Every ton of ash reused reduces waste and saves emissions associated with the production of new materials. On a global scale, if current trends continue, biochar and ash can play a significant role in achieving the goals of the Paris Agreement by reducing emissions and increasing carbon dioxide absorption. There is already a growing interest in these solutions around the world, from pilot projects on farms in Africa, to commercial biochar installations in Europe and North America, to government programs in Asia. With scientific support and appropriate regulations, the use of ash and biochar can become one of the pillars of sustainable development, bringing measurable environmental benefits to current and future generations.

References

1. Z.H. Weng, A.L. Cowie, Estimates vary but credible evidence points to gigaton-scale climate change mitigation potential of biochar. Commun. Earth Environ. **6**(1), 1–9 (2025). https://doi.org/10.1038/S43247-025-02228-X
2. V. Viola, M. Catauro, A. D'Amore, P. Perumal, Assessing the carbonation potential of wood ash for CO_2 sequestration. Low-Carbon Mater. Green Constr. **2**(1), 1–14 (2024). https://doi.org/10.1007/S44242-024-00043-9/FIGURES/13
3. S. Jung, Y.K. Park, E.E. Kwon, Strategic use of biochar for CO_2 capture and sequestration. J. CO_2 Util. **32**, 128–139 (2019). https://doi.org/10.1016/J.JCOU.2019.04.012
4. M. Karimi, M. Shirzad, J.A.C. Silva, A.E. Rodrigues, Biomass/biochar carbon materials for CO_2 capture and sequestration by cyclic adsorption processes: a review and prospects for future directions. J. CO_2 Util. **57**, 101890 (2022). https://doi.org/10.1016/J.JCOU.2022.101890
5. L. Wang, D. Chen, L. Zhu, Biochar carbon sequestration potential rectification in soils: synthesis effects of biochar on soil CO_2, CH_4 and N_2O emissions. Sci. Total Environ. **904**, 167047 (2023). https://doi.org/10.1016/J.SCITOTENV.2023.167047
6. T.F. Rittl et al., High application rates of biochar to mitigate N_2O emissions from a N-fertilized tropical soil under warming conditions. Front. Environ. Sci. **8**, 611873 (2021). https://doi.org/10.3389/FENVS.2020.611873/BIBTEX
7. P. Grutzmacher, A.P. Puga, M.P.S. Bibar, A.R. Coscione, A.P. Packer, C.A. de Andrade, Carbon stability and mitigation of fertilizer induced N_2O emissions in soil amended with biochar. Sci. Total Environ. **625**, 1459–1466 (2018). https://doi.org/10.1016/J.SCITOTENV.2017.12.196
8. H. Wang et al., Biochar enhances paddy productivity, carbon sequestration, and reduces greenhouse gas emissions in the middle Yangtze River region. Agronomy **14**(12), 3067 (2024). https://doi.org/10.3390/AGRONOMY14123067/S1
9. F. Wang, D. Ouyang, Z. Zhou, S.J. Page, D. Liu, X. Zhao, Lignocellulosic biomass as sustainable feedstock and materials for power generation and energy storage. J. Energy Chem. **57**, 247–280 (2021). https://doi.org/10.1016/J.JECHEM.2020.08.060

10. Y. Zhang et al., A review of biomass pyrolysis gas: forming mechanisms, influencing parameters, and product application upgrades. Fuel **347**, 128461 (2023). https://doi.org/10.1016/J.FUEL.2023.128461
11. H. Stančin, V. Strezov, H. Mikulčić, Life cycle assessment of alternative fuel production by co-pyrolysis of waste biomass and plastics. J. Clean. Prod. **414**, 137676 (2023). https://doi.org/10.1016/J.JCLEPRO.2023.137676
12. B. Carrasco, N. Cruz, J. Terrados, F.A. Corpas, L. Pérez, An evaluation of bottom ash from plant biomass as a replacement for cement in building blocks. Fuel **118**, 272–280 (2014). https://doi.org/10.1016/J.FUEL.2013.10.077
13. X. Lu et al., Recycling of coal fly ash in building materials: a review. Minerals **13**(1), 25 (2022). https://doi.org/10.3390/MIN13010025
14. N. Kumar, K. Kupwade-Patil, R. Higuchi, D.P. Ferrell, V.A. Luttrull, J.G. Lynam, Use of biomass ash for development of engineered cementitious binders. ACS Sustain. Chem. Eng. **6**(10), 13122–13130 (2018). https://doi.org/10.1021/ACSSUSCHEMENG.8B02657
15. J. Khunthongkeaw, S. Tangtermsirikul, T. Leelawat, A study on carbonation depth prediction for fly ash concrete. Constr. Build. Mater. **20**(9), 744–753 (2006). https://doi.org/10.1016/J.CONBUILDMAT.2005.01.052
16. L. Baxter, Biomass-coal co-combustion: opportunity for affordable renewable energy. Fuel **84**(10), 1295–1302 (2005). https://doi.org/10.1016/J.FUEL.2004.09.023
17. S.G. Sahu, N. Chakraborty, P. Sarkar, Coal–biomass co-combustion: an overview. Renew. Sustain. Energy Rev. **39**, 575–586 (2014). https://doi.org/10.1016/J.RSER.2014.07.106
18. E.S. Hope, D.W. McKenney, D.J. Allen, J.H. Pedlar, A cost analysis of bioenergy-generated ash disposal options in Canada. Can. J. For. Res. **47**(9), 1222–1231 (2017). https://doi.org/10.1139/CJFR-2016-0524
19. S. Donatello, M. Tyrer, C.R. Cheeseman, EU landfill waste acceptance criteria and EU hazardous waste directive compliance testing of incinerated sewage sludge ash. Waste Manage. **30**(1), 63–71 (2010). https://doi.org/10.1016/J.WASMAN.2009.09.028
20. J.B. Shah, J.B. Valaki, A comprehensive review on feasibility of different agro residues for production of bio-oil, bio-char and pyro-gas. J. Kejuruter. **35**(1), 77–93 (2023). https://doi.org/10.17576/JKUKM-2023-35(1)-08
21. S. Marinos, A. Agapios, P. Terpsithea, T. Michail, A.A. Zorpas, H. Hamdi, Biochar production from the pyrolysis of tomato processing residues, in *Tomato Processing By-Products: Sustainable Applications*, Jan 2022, pp. 171–200. https://doi.org/10.1016/B978-0-12-822866-1.00002-8
22. A. Maresca, J. Hyks, T.F. Astrup, Recirculation of biomass ashes onto forest soils: ash composition, mineralogy and leaching properties. Waste Manage. **70**, 127–138 (2017). https://doi.org/10.1016/j.wasman.2017.09.008
23. Z. Tan, A. Lagerkvist, Phosphorus recovery from the biomass ash: a review. Renew. Sustain. Energy Rev. **15**(8), 3588–3602 (2011). https://doi.org/10.1016/J.RSER.2011.05.016
24. B. Symanowicz, M. Becher, D. Jaremko, K. Skwarek, Possibilities for the use of wood ashes in agriculture. J. Ecol. Eng. **19**(3), 191–196 (2018). https://doi.org/10.12911/22998993/86156
25. G. Ondrasek et al., Biomass bottom ash & dolomite similarly ameliorate an acidic low-nutrient soil, improve phytonutrition and growth, but increase Cd accumulation in radish. Sci. Total Environ. **753**, 141902 (2021). https://doi.org/10.1016/j.scitotenv.2020.141902
26. J.I. Odzijewicz, E. Wołejko, U. Wydro, M. Wasil, A. Jabłońska-Trypuć, Utilization of ashes from biomass combustion. Energies **15**(24), 9653 (2022). https://doi.org/10.3390/EN15249653
27. Y. Tao et al., Utilization of cotton byproduct-derived biochar: a review on soil remediation and carbon sequestration. Environ. Sci. Eur. **36**(1), 1–16 (2024). https://doi.org/10.1186/S12302-024-00908-7/METRICS
28. M. Pazalja et al., Heavy metals content in ashes of wood pellets and the health risk assessment related to their presence in the environment. Sci. Rep. **11**(1), 1–9 (2021). https://doi.org/10.1038/S41598-021-97305-4
29. X. Yao et al., Characterization comparison of bottom ash and fly ash during gasification of agricultural residues at an industrial-scale gasification plant—experiments and analysis. Fuel **285**, 119122 (2021). https://doi.org/10.1016/J.FUEL.2020.119122

30. C. Yao et al., Formation mechanisms and degradation methods of polycyclic aromatic hydrocarbons in biochar: a review. J. Environ. Manage. **357**, 120610 (2024). https://doi.org/10.1016/J.JENVMAN.2024.120610
31. I. Maj, K. Niesporek, P. Płaza, J. Maier, P. Łój, Biomass ash: a review of chemical compositions and management trends. Sustainability **17**(11), 4925 (2025). https://doi.org/10.3390/SU17114925
32. R. Martínez-García et al., The present state of the use of waste wood ash as an eco-efficient construction material: a review. Materials **15**(15), 5349 (2022). https://doi.org/10.3390/MA15155349
33. Q. Liu et al., Effects of modified biochars on the shifts of short-chain fatty acid profile, iron reduction, and bacterial community in paddy soil. FEMS Microbiol. Ecol. **98**(12) (2022). https://doi.org/10.1093/FEMSEC/FIAC131
34. L. Genesio, F. Miglietta, S. Baronti, F.P. Vaccari, Biochar increases vineyard productivity without affecting grape quality: results from a four years field experiment in Tuscany. Agric. Ecosyst. Environ. **201**, 20–25 (2015). https://doi.org/10.1016/J.AGEE.2014.11.021
35. K.N. Shoudho, T.H. Khan, U.R. Ara, M.R. Khan, Z.B.Z. Shawon, M.E. Hoque, Biochar in global carbon cycle: towards sustainable development goals. Curr. Res. Green Sustain. Chem. **8**, 100409 (2024). https://doi.org/10.1016/J.CRGSC.2024.100409
36. *Global Biochar Market Soars to $600 Million in 2023, Setting the Stage for Future Growth—International Biochar Initiative*. Accessed 30 June 2025. [Online]. Available: https://biochar-international.org/news/global-biochar-market-soars-to-600-million-in-2023-setting-the-stage-for-future-growth/
37. E. Menéndez, C. Argiz, H. Recino, M.Á. Sanjuán, Characterization of mortars made with coal ashes identified as a way forward to mitigate climate change. Crystals **12**(4), 557 (2022). https://doi.org/10.3390/CRYST12040557
38. I. Maj, S. Kalisz, S. Ciukaj, Properties of animal-origin ash—a valuable material for circular economy. Energies **15**(4), 1274 (2022). https://doi.org/10.3390/EN15041274
39. M. Marcińczyk, Y.S. Ok, P. Oleszczuk, From waste to fertilizer: nutrient recovery from wastewater by pristine and engineered biochars. Chemosphere **306**, 135310 (2022). https://doi.org/10.1016/J.CHEMOSPHERE.2022.135310
40. A. Zuševica, S. Celma, S. Neimane, M. von Cossel, D. Lazdina, Wood-ash fertiliser and distance from drainage ditch affect the succession and biodiversity of vascular plant species in tree plantings on marginal organic soil. Agronomy **12**(2), 421 (2022). https://doi.org/10.3390/AGRONOMY12020421
41. L. Sujeeun, S.C. Thomas, Biochar rescues native trees in the biodiversity hotspot of Mauritius. Forests **13**(2), 277 (2022). https://doi.org/10.3390/F13020277/S1

Chapter 9
Conclusion

Biomass ash and biochar emerge from the present analysis as promising and sustainable soil amendments that can partially substitute for conventional mineral fertilizers. Biomass ash, generated as a residue from the combustion of woody and herbaceous biomass, contains significant amounts of macronutrients such as calcium, potassium, phosphorus, and magnesium. Its alkaline nature enables effective neutralization of acidic soils. For this reason, biomass ash is often considered functionally comparable to liming agents. In nutrient-poor soils, it provides essential elements that are readily available to plants. Biochar, by contrast, is produced through the pyrolysis of biomass under limited oxygen conditions. It is characterized by a high proportion of stable organic carbon that resists biological degradation, making it an efficient long-term carbon sink in soils, potentially lasting hundreds or even thousands of years. Its porous structure and large surface area enhance the soil's cation exchange capacity and water retention potential, thereby improving fertility and reducing nutrient losses from the soil profile.

The scientific literature of the past decade increasingly highlights that biochar application can improve crop yields and reduce agricultural greenhouse gas emissions. These effects place biochar within the broader framework of climate-smart agriculture. Similarly, the utilization of biomass ash represents a circular strategy whereby nutrients initially taken up by plants are returned to the soil in a usable form rather than being discarded as waste.

Both materials significantly influence soil chemical and physical properties. Ash and biochar typically lead to increased soil pH, higher base saturation, and improved soil structure. Biomass ash provides base-forming cations and reduces soil acidity, thereby enhancing nutrient retention and availability. Biochar contributes stable organic carbon, which supports a range of beneficial changes including increased cation exchange capacity, improved water-holding capacity, and formation of soil aggregates. Experimental data confirm that biochar can improve the structural stability and hydrological functions of soils while stimulating microbial activity. Additionally, biochar is effective at stabilizing soil nitrogen by reducing ammonia

K. Chojnacka et al., *Sustainable Fertilizers: Utilizing Biomass Ashes and Biochars*, SpringerBriefs in Applied Sciences and Technology,
https://doi.org/10.1007/978-3-032-03765-7_9

volatilization and nitrate leaching. This stabilization effect can also contribute to a reduction in nitrous oxide emissions, a potent greenhouse gas. These mechanisms explain why both ash and biochar positively affect plant growth and overall soil functioning.

The agricultural use of biomass ash and biochar generates multiple environmental co-benefits. Biochar, in particular, acts as a permanent carbon sink. Incorporation of biochar into soils removes atmospheric CO_2 and stores it in a stable form, contributing to global carbon sequestration efforts. Enriching soils with both materials also reduces emissions of other greenhouse gases from agricultural systems and diminishes reliance on synthetic fertilizers that are energy-intensive to produce. Replacing a portion of mineral fertilizers with biomass ash can therefore reduce energy consumption and lower the carbon footprint associated with fertilizer manufacturing. Research further indicates that co-application of ash and organic amendments improves plant health, enhances chlorophyll content, and stimulates soil enzyme activities. These changes support a reduction in the need for chemical pesticides. In this way, bio-based products serve not only as nutrient sources but also as enhancers of soil ecological resilience. The recycling of biomass ash represents a clear manifestation of circular economy principles. Nutrients absorbed by plants during growth are reintroduced into the soil in a closed-loop system, preventing long-term nutrient depletion. An additional environmental benefit lies in diverting ash from landfills, transforming it from a problematic by-product into a valuable agricultural input.

Despite these advantages, the use of biomass ash and biochar is associated with certain limitations and risks. The chemical composition of ash is highly variable and depends on the type and origin of the biomass feedstock, as well as combustion conditions. As a result, both the fertilizing value and potential environmental risks may vary between batches. Particular attention must be paid to contaminants such as heavy metals including cadmium, lead, nickel and mercury, and organic toxicants such as dioxins. These substances may be present in ash derived from contaminated feedstocks or from poorly controlled combustion processes. If the concentrations of such pollutants exceed legal thresholds, the use of ash in agriculture may pose serious environmental and health hazards. However, studies show that many biomass ashes, especially those obtained from the combustion of clean wood or straw, meet regulatory safety standards for fertilizers. Even so, regular quality monitoring and compliance with established norms remain essential to minimize ecological risk.

The properties of biochar also vary considerably depending on feedstock type and pyrolysis parameters. This variability poses a challenge, as different soil types and crops require tailored biochar formulations with defined physicochemical characteristics. Therefore, it is necessary to develop and enforce quality standards for biochar, including benchmarks for carbon content, ash proportion and porosity, to ensure predictable performance under field conditions. Furthermore, both ash and biochar have nutritional limitations. Biomass ash contains negligible nitrogen, since nitrogen is almost entirely lost during combustion in the form of gaseous emissions. Consequently, although ash supplies valuable minerals, it cannot replace nitrogen fertilizers. In agricultural use, it should therefore be supplemented with nitrogen from other sources to meet plant requirements. Biochar also contains only limited nitrogen, although it contributes to nitrogen retention in the soil matrix. Most of the studies analyzed in this volume were short- to medium-term, and knowledge regarding the long-term cumulative effects of repeated application of these materials in various agroecosystems is still limited.

The results presented in this book demonstrate the significant potential of biomass ash and biochar technologies to support the transition toward more sustainable and climate-resilient agriculture. Based on the available evidence, the following key recommendations can be formulated:

- Continued long-term field studies on the effects of biomass ash and biochar on soils, crop performance and environmental quality are essential. These studies will improve understanding of their temporal impacts and help optimize application frequency, incorporation methods and material combinations. Regular monitoring of treated soils for changes in heavy metal content and physicochemical properties is also recommended.
- Harmonization of quality standards for both ash and biochar intended for agricultural use is crucial. Several European countries have already introduced legal thresholds for contaminants and minimum quality parameters. Such regulations should be adopted more broadly. In addition, legal frameworks should be revised to facilitate the classification of safe ashes as fertilizer products rather than as waste. A clear system of certification would increase farmer confidence and encourage wider adoption.
- Biochar and ash should be integrated into broader soil fertility management strategies. Co-application with organic materials such as compost, manure, or digestate can balance nutrient supply, particularly nitrogen, and enrich biochar with additional fertilizing properties. This synergistic approach has proven effective in improving nutrient use efficiency. It is also advisable to include both materials in soil remediation programs, where their liming and sorptive properties can aid in the restoration of degraded lands.
- Awareness-raising efforts targeted at farmers, agronomists and policymakers are needed to communicate the agronomic and environmental benefits of biomass ash and biochar. On-farm demonstrations, training sessions, and the publication of technical guidelines can support the implementation of best practices. Economic incentives such as subsidies, tax relief, or carbon credits could further accelerate adoption. Biochar in particular has been officially recognized by the Intergovernmental Panel on Climate Change (IPCC) as a negative emissions technology suitable for carbon offset programs, opening the door to market-based mechanisms that reward its use.

In conclusion, the findings presented in this volume confirm the substantial potential of biomass ash and biochar as innovative soil amendments and sustainable fertilizer alternatives. When applied appropriately and with due regard for quality control, these materials can support yield stability, restore soil health, reduce environmental impacts of farming and contribute to the efficient management of organic waste streams. Further technological development, combined with regulatory support and practical guidance, will be critical to scaling these innovations. Nonetheless, current evidence already points to a central role for ash and biochar in the future of agriculture that balances productivity with environmental stewardship.

Acknowledgements Funded by the European Union under the HORIZON Coordination and Support Actions project no. 101181936—FertiCovery: Best available techniques to recover or recycle fertilising products from secondary raw materials.

Statements and Declarations

Competing Interests The authors have no competing interests to declare that are relevant to the content of this article.

Ethical Approval and Consent to Participate Not applicable.

Author Contributions **Katarzyna Chojnacka**—Conceptualization, Supervision, Writing—Original Draft; Writing—Review and Editing; **Dawid Skrzypczak**—Writing—Original Draft; **Grzegorz Izydorczyk**—Writing—Original Draft; **Filip Gil**—Writing—Original Draft.

FSC
www.fsc.org
MIX
Papier aus verantwortungsvollen Quellen
Paper from responsible sources
FSC® C105338

If you have any concerns about our products,
you can contact us on
ProductSafety@springernature.com

In case Publisher is established outside the EU,
the EU authorized representative is:
Springer Nature Customer Service Center GmbH
Europaplatz 3, 69115 Heidelberg, Germany

Printed by Libri Plureos GmbH
in Hamburg, Germany